# Ecology, Ethics and Hope

# Ecology, Ethics and Hope

Edited by Andrew T. Brei

Published by Rowman & Littlefield International, Ltd.
Unit A, Whitacre Mews, 26-34 Stannary Street, London SE11 4AB
www.rowmaninternational.com

Rowman & Littlefield International, Ltd. is an affiliate of Rowman & Littlefield
4501 Forbes Boulevard, Suite 200, Lanham, Maryland 20706, USA
With additional offices in Boulder, New York, Toronto (Canada), and London (UK)
www.rowman.com



**British Library Cataloguing in Publication Information Available**
A catalogue record for this book is available from the British Library

ISBN: HB 978-1-78348-549-9
ISBN: PB 978-1-78348-550-5

**Library of Congress Cataloging-in-Publication Data**

Ecology, ethics, and hope / edited by Andrew T. Brei.
pages cm.
Includes bibliographical references and index.
ISBN 978-1-78348-549-9 (cloth : alk. paper) -- ISBN 978-1-78348-550-5 (pbk. : alk. paper) -- ISBN 978-1-78348-551-2 (electronic)
1. Hope. 2. Global environmental change. 3. Environmental ethics. 4. Climatic changes--Moral and ethical aspects. I. Brei, Andrew T., 1974- editor.
BD216.E26 2015
179'.1--dc23
2015030890

∞™ The paper used in this publication meets the minimum requirements of American National Standard for Information Sciences Permanence of Paper for Printed Library Materials, ANSI/NISO Z39.48-1992.

Printed in the United States of America

Dedicated to Ella and Mia, my two brightest hopes

# Contents

# Introduction

In general, despair is a bad thing. Suffering in its grip, some act in ways that are destructive, while others are stunned to the point of inaction. The loss of hope can be sad and scary, and it prompted Elif Şhafak (2010) to say, '[w]hatever happens in your life, no matter how troubling things might seem, do not enter the neighborhood of despair' (73).

To combat despair, many grasp at hope. And among contemporary philosophers there are some who have started wondering about what this turn toward hope entails.[1] This book is a collection of works about hope. Most of these works were written especially for this volume, and all were written by authors who agree that despair is particularly bad when it either exacerbates ecological crises or allows them to go unchecked. These authors agree that there are things we can do—things we ought to do—in order to avoid the kinds of destruction and inaction that often result from despairing.

The divergence in the pieces that follow has to do with exactly *what* we can and ought to do. Some will focus on identifying the sources of despair and on suggesting remedies that can engender hope and inspire action. Andre and Brei do this in part I. The sources of despair they identify include grim scientific reports of escalating problems, a culture that promotes naïve optimism over realism and the overwhelming complexity of environmental problems. The remedies include acknowledging the effects that negative messages can have on the psyche, finding ways to merge honesty with hope, as well as revealing the extent to which hope that is the product of social pressure is stripped of its motivational power. Others—Fiala, Nolt and Hedberg—examine the moral imperative that exists regarding hope, arguing (on various grounds) that we have a duty to be hopeful. This duty may in turn oblige us to value and treat the natural world in ways that would substantially curb the damage humans currently do to it. Such an approach is taken up in part II.

In part III, the focus shifts to hope as a virtue. According to Treanor, hope ought to be understood as an environmental virtue, alongside simplicity and political engagement. Hope is the foil to despair (a vice) and is necessary for human flourishing. Beever agrees that hope is a virtue as well as an important component of ethical decision-making—but to say that hope is necessary is problematic. The sheer complexity and fluidity of environmental crises stymies hope's ability to fixate on a target, making it difficult to envision and realize outcomes. Best, says Beever, to let hope merely serve in a practical capacity as an indicator of environmental problems worth dealing with.

Finally, in part IV, Nelson and Kretz deal with hope and its relation to motivation. Nelson is highly sceptical of hope's ability to bring about positive changes because, among other things, of its preoccupation with consequences. He suggests that we would be far better off if we relied upon reasons and values that are not so easily dashed—reasons and values that center upon care and generosity and other basic virtues. Kretz, on the other hand, sings hope's praise, pointing to the psychological and social benefits it provides that could go a long way toward reversing negative ecological trends. The value of hope can be reflected in its importance for individual as well as collective (global) good.

Despite the differences in perspective and prescription, these pieces converge in a number of places. Again, all of the chapters in this book were written by people who regard environmental despair with dismay. Most of the authors in this collection agree that hope has the power to make positive changes to one's mind and attitude—the kinds of changes that can find expression through individual, community and global action. Thus, there are clear psychological and psychosocial threads running throughout the pieces in this collection. In addition, there is unity around the idea that any sort of hope worth discussing (or having) has to be understood as more than optimism or desire. Hope requires justification. Thus, there is a common interest in stating clearly what hope amounts to. And running throughout the book is the acknowledgement that whatever else we ought to say about it, hope is clearly an important component in environmental discussions today.

## NOTES

1. Hope has received more attention over the last several decades from psychologists and healthcare practitioners (who are interested in hope's function as well as its relation to other emotions/attitudes) than it has from philosophers. So I am glad to see more attention being paid to hope by philosophers—especially those interested in ecological issues.

*Part I*

# Hope and Despair

*Chapter One*

# The Need to Talk About Despair

Elizabeth Andre

As scientists become more and more certain of anthropogenic climate change, as their projections become more dire, as their warnings become more urgent and as climate-related disasters become more frequent and spectacular, one might think the public would start to share in the concern and start to demand action. Belief in anthropogenic climate change, however, remains spotty and the issue consistently fails to rise to the top of lists of public concerns (Riffkin 2014). This paradox confounds us environmentalists, who wonder what could be more urgent than the potential undermining of the ability of our planet to support life as we know it. Why, we wonder, have our scientific reports, documentaries, books and magazine articles, marches, music festivals and art installations failed to move people to action? We proffer numerous explanations, including the politicized nature of the climate issue, America's scientific illiteracy, anti-science and anti-intellectual sentiment, the denial industry funded by the fossil fuel lobby, financial pressures on families, a dearth of empathy with others who will be more vulnerable than we and other moral shortcomings. We examine our messaging to see if there is a better word for doom (Montenegro 2009).

Could it be, however, that none of the 'causes' of disengagement mentioned above are actually causal? What if they are merely convenient 'outs' for people who, at the most basic level, do not want to be depressed and frightened? We environmentalists seldom acknowledge the effects our messages can have on our (and others') emotions. We have yet to explicitly grapple with the reasons we feel despair. We have yet to create safe spaces for sharing fears and doubts. We have yet to provide adequate tools for transforming those negative emotions into hope.

Perhaps we can learn from ethnographer Kari Marie Norgaard's study in Norway where, although many of the factors blamed for the backlash against climate science in the United States are either absent or much reduced, Norgaard (2006) concluded that Norwegians still avoid talking about climate change because of the unpleasant feelings it produces, including powerlessness, fear and guilt. We must discover how to deal with these negative emotions before our work on any of the aforementioned 'causes' of disengagement will yield results.

If we environmentalists agree with Aldo Leopold that we are like 'the doctor who sees the marks of death in a community that believes itself well and does not want to be told otherwise', then we need to examine the emotional toll exacted upon doctors of terminally ill patients and acknowledge that we too experience these emotions (Leopold 1953, 165). We need to recognize, as does physician and medical school professor Rachel Remen (1997), that we must prepare our students to deal with their own grief in witnessing loss and in being unable to prevent death.

If we agree with environmental activist Yvon Chouinard[1] that environmentalism is similar to a religion in that we are always striving to become more enlightened and faithful, then we need to recognize the effects of religious stress and doubt. We need to recognize, as does religious scholar Uzeyir Ok, that all moral education systems must help students deal with 'uncertainty, distress, questioning, contradiction, confusion, doubt, anxiety, loss of faith, . . . guilt, alienation, loneliness, disloyalty, and unwillingness' (Ok 2004, 201).

## SOURCES OF ENVIRONMENTAL DESPAIR

American culture tends to value optimism and to see pessimism as a personal failing (Held 2002). From the reactions of others, people quickly learn to not bring up 'depressing' topics like climate change in 'polite company' for fear of seeming like a Debbie Downer. These social norms isolate those who feel despair and can lead them to use emotional management strategies including denial and distancing from the issue. If we can understand the sources of environmental despair and realize that these feelings do not come from personal weakness, we can begin to create spaces to talk openly about these emotions and begin to create more productive reactions. Sources of environmental despair include the complexity of issues, the intersection of our natural biophilia with destruction and loss, tactics of environmental education, the dominant environmental paradigms and a culture of pessimism within intellectual circles. I'll deal with each issue below.

Let's assume for now that hope is related to a decision to act (we'll explore this later). Research in environmental education suggests that, before a person decides to take action on an environmental issue, he or she must have in-depth knowledge of the issues, knowledge of and skill in using environmental action strategies and a belief that his or her actions can make a difference. These are labelled ownership and empowerment variables (Hungerford & Volk 1990). With a relatively simple environmental problem, such as point-source pollution in a local stream, these prerequisites are easily met. With modern global environmental challenges such as climate change, however, this is a high bar. An in-depth knowledge of the issue would require study not only in climate science, but also in conservation ecology, international and domestic politics, economics, demographics, agricultural science and energy production. The complexity of this quintessential collective action problem makes it difficult to see how one's individual actions can make a difference (Abbasi 2005). With extremely complex issues, we are often left without feelings of ownership and empowerment.

A second source of environmental despair is the intersection of environmental loss and destruction with our natural biophilia, defined as our innate affinity with the natural world and with other living beings (Wilson 1986). For example, whose soul doesn't thrill to an episode of the BBC's *Planet Earth*? And, after watching, who doesn't feel a twinge of pain at seeing the headline, 'Seven years since the *Planet Earth* franchise first started production, has the footage become a chronicle of an already vanished world?' (Kaufman 2009). We love other species, and it is painful to see them suffering and dying.

The field of environmental education knows the emotional power of causing someone to love a place and then showing its destruction—research shows it can be a catalyst for turning people into activists (Chawla & Cushing 2009). Environmental educators purposely design materials to open people's hearts and then deliver an emotional hit. We don't know, however, how frequently we can 'hit' people's hearts before they turn away. We may, unwittingly, be contributing to the same disengagement we are working to overcome.

Environmental despair also comes from the two dominant paradigms in environmental thinking: Eco-Marxism and Ecological Modernization. They each can lead to despair, although in different ways. Eco-Marxism argues that capitalism and its reliance on growth is fundamentally incompatible with our finite planet, that we are all on a treadmill of production and are unwilling or unable to get off and that capitalism cannot be reformed. This argument is logical and seems to be supported by history. Derrick Jensen, and other thinkers in this school, resonate with many students who come to believe that, for environmental salvation, 'civilization needs to be brought down now' (Jensen 2006b, xi). As those students, believing the only way forward is complete revolution, are confronted with the reality that change is a difficult and slow process, they can experience apathy, outrage, anger, hopelessness and pessimism (Sheppard 2004, 216).

Ecological Modernization, which on the surface seems much more hopeful, argues that capitalism can be greened and that innovation and technology will allow us to sustainably continue our civilization. Thinkers in this school, such as Michael Braungart and William McDonough, authors of *Cradle to Cradle* (2002), argue that civilization is the answer to our problems, not the source, and that we can learn from the mistakes of the industrial revolution to design industries that are in harmony with natural systems. How could this seemingly hopeful paradigm contribute to despair? We will have to return to this question once we have explored the nature of hope.

One final source (for now) of environmental despair is a culture within intellectual circles where pessimism is seen as sophisticated and realistic, while hope is considered 'quaint, religious, and naïve' (Johnson 2005, 48). This view is not without merit, as pessimists are often more accurate (Norem & Chang 2002), and the dire assessment of our situation provided by Derrick Jensen may be closer to the truth than that of William McDonough (Shaw 2006). Before we can make judgements about the value of hope, however, we must understand what it is and is not.

## WHAT IS HOPE?

Although these words are sometimes used interchangeably, we must distinguish optimism from hope. Optimism is belief-centred and irrational. For example, in the case of climate change, to believe that before the end of this century we will stabilize atmospheric concentrations of carbon dioxide below 350 ppm (the level scientists warn us we must achieve if we want to 'maintain a planet similar to the one on which civilization developed and to which life on Earth is adapted' [Hansen et al. 2008, 217]) would be naïve, demonstrating a lack of understanding of the problem and ignorance of the best predictive models (Anderson & Bows 2008). To truly expect this to happen would be *optimism* and would, indeed, be quaint.

Environmentalist Bill McKibben, founder of the organization 350.org, however, has this as his goal. Far from being naïve, McKibben deeply understands the issue and the scope of the challenge. He has chosen this goal because it is the only one that makes sense to him (why choose a goal that would do anything *other* than maintain a planet similar to the one on which civilization developed and to which life on Earth is adapted?), not because he expects to achieve it. In this sense, rather than being optimism, this is hope. We find a definition of hope, along with insights into this seeming paradox from philosophy, cognitive psychology and medical literature.

Philosopher David Orr, when contemplating the environmental holocaust, distinguishes between optimism and hope. He admonishes us that there is:

> not one speck of ground for wishful thinking of any kind . . . but hope can be nurtured by good work, openness to life, and rising above our lesser selves. Hope, real hope, comes from doing the things before us that need to be done in the spirit of thankfulness and celebration, without worrying about whether we will win or lose. (Orr 2004, 297)

McKibben must know, intellectually, that he will likely not win, if winning is defined as narrowly as I have defined it above. He is, however, 'doing the things that need to be done in the spirit of thankfulness and celebration'. This strategy makes more sense if we reject the dominant narrative of tragedy and its metaphor of warfare, with its false dualism and the goal of absolute victory through destruction of an enemy. Instead we frame our endeavours as a comedy, with its metaphor of a game, and its goal of supporting community (Meeker 2004, 297). From a very practical perspective, this view allows us to start working toward ameliorating the problem, even when we do not see a path to absolute victory.

Cognitive psychologists, within Hope Theory, perceive hope as a cognitive process of working toward a goal that is realistically achievable. They hypothesize that the human brain evolved to develop strategies to create pathways to goals, and that the brain is happiest when it is actively working to move along those pathways. This would explain why, often, when a person achieves suicide, his family and friends are surprised because, although they had been worried about the person in the past, in the weeks leading up to the suicide, the person had seemed so much happier. Hope Theorists would explain that, once the person had a plan (to kill himself), his brain became engaged in working out the steps to achieving that goal, and the process of planning created good emotions and, paradoxically, hope.

For a goal to be appropriate for developing hope, it must be challenging enough to absorb one's attention, but not so challenging as to be impossible to achieve. When people's chosen goals are so difficult to achieve that they continually are thwarted in their efforts to make progress, they often experience negative emotions (Snyder 2002). At first glance, it may seem that Bill McKibben's goal is not appropriate for Hope Theory, as it is arguably impossible to achieve. Because, however, McKibben is actively working on smaller goals (for example, by choosing the battle against the Keystone XL Pipeline as a rallying issue), with the end goal of climate stabilization at 350 ppm as merely a guiding compass, he is able to make progress toward his smaller goals. Although smaller than the ultimate goal of climate stabilization at 350 ppm, McKibben's intermediate goals are still ambitious enough to completely absorb his attention and to fully engage his brain in planning and working. Hope Theorist C. R. Snyder finds insights from devout Christians whose ultimate goal is to achieve the perfection of Christ. Even though they believe this goal is impossible to achieve, it guides their choice of smaller goals, giving them enjoyment in the journey (Snyder 2002, 266). This process of active planning and working is the essence of hope, according to Hope Theory. This process itself, and the positive emotions produced through progress on these smaller goals, provides the antidote to creeping disempowerment, despair, guilt and other negative emotions.

Cognitive psychology's definition of hope supports the philosophical stance of David Orr. Both perspectives distinguish between naïve optimism and true hope, which is based in action. Both understandings allow for us to comprehend the severity and complexity of the issue as well as the unlikeliness of a narrowly defined success, and yet still find ownership and empowerment.

## OTHER SOURCES OF ENVIRONMENTAL DESPAIR

Now that we understand the difference between optimism and hope, we can explore several more sources of environmental despair including false hope, utopian visions and the ecological modernization paradigm. Hope Theory gives us insights into false hope, defined as having a goal and the requisite motivation to work toward it, but lacking an appropriate plan to achieve that goal (Kwon 2002). Unwitting environmentalists and environmental educators sometimes push audiences toward false hope by impressing upon them the enormity and import of the issue and then failing to show how people can begin to successfully work toward alleviating the issue. For example, Al Gore's 2007 film *An Inconvenient Truth* spends the majority of its one hundred minutes painting a picture of a future of climate chaos and societal collapse and then, almost as an afterthought, overlays the final credits with about three and a half minutes of suggestions for action, many of which are individually focused, such as changing lightbulbs, buying energy-efficient appliances, weatherizing your home, buying a hybrid vehicle and taking public transportation or bicycling. After spending the previous ninety-six minutes coming to understand the complexity of this quintessential collective-action problem and seeing the inertia in its key drivers, the audience knows intuitively that these individual actions are insufficient. Many of the other suggestions are platitudes, such as 'tell your parents not to ruin the world you will live in', are out of reach for most people, such as 'run for congress', or are inaccessible without specialized expertise, such as 'reduce our dependence on foreign oil' and 'help farmers grow alcohol fuels'. The end result for many who watch *An Inconvenient Truth* is false hope: a goal (to stop climate change), the motivation to do it and a lack of an appropriate path to achieve that goal. Eventually they must take stock of their effectiveness toward achieving their goal. Lack of progress can produce negative feelings, including the guilt, fear and helplessness that Norgaard found led people to use emotional management strategies including denial and distancing.

Closely related to false hope are utopian visions—common yet counter-productive motivational tools used by environmentalists. Geographer David Pepper critiques various radical environmental paradigms, including ecotopianism, deep ecology and bioregionalism, which use a beautiful vision of the future as a tool for inspiring people to action. Pepper finds these utopian paradigms to be more reactionary fantasies than potentially transgressive or transformative movements. Pepper asserts for the paradigms to be truly transgressive, they 'must be rooted in existing social and economic relations rather than being merely a form of abstraction unrelated to the processes and situations operating in today's "real" world' (Pepper 2005, 18). If the vision is indeed a fantasy, people either can't figure out how to get there, so they don't start taking any action, or if they do try to take action, they experience unrealistic expectations of what will happen as a result of their work. When expectations are not met, they often end up burning out (Breyman 1997, 5).

Finally, we will revisit ecological modernization as a source of environmental despair. While it may not promote pessimism, as does the competing paradigm of Eco-Marxism, it encourages an optimistic stance, rather than a hopeful one. As discussed earlier, ecological modernization argues that, through the process of the free market, capitalism will reform itself through technological innovations to be more sustainable. Setting aside the many critiques of the particulars of ecological modernization, including arguments by policy experts like Joseph Romm, who argue that the very idea of a 'technological breakthrough' is a myth (Romm 2007), we will examine only the paradigm's relationship to hope, as understood by both Orr and Hope Theory. As seen earlier, hope requires the individual to be an *agent* in the process of actively moving toward a goal. Ecological modernization, however, removes the power for action from the typical individual (and from collectives of individuals, including government) and relies, instead, on technology and on the market (Lueck 2007, 257). This paradigm minimizes the need for individual and community action to change the underlying structures of the economy, because it sees the structure of capitalism as central to allowing a sustainable solution to evolve. Individuals, believing the goal can be achieved without their participation, are left without the ownership, empowerment and action that create hope. As they wait for the deus ex machina to save society, the media continues to report on the worsening situation.

We've touched on many of the sources of environmental despair, including the complexity of modern environmental issues, the intersection of our natural biophilia with images of destruction and loss, tactics of environmental education that prey on our environmental sensitivities, a culture of pessimism within intellectual circles that dismisses hope as quaint, utopian visions that are devoid of social change potential, the creation of false hope through a failure to provide appropriate pathways to goals and dominant environmental paradigms that promote either pessimism or optimism, but not true hope. We've also discussed how environmental despair can lead people to use emotional management strategies including denial and distancing from the issue and how, if we environmentalists are to move people to action, we will need to provide them with not only an understanding of the sources of their despair that does not frame it as a personal failing, but also with a toolbox of strategies for creating true hope.

## HOW DO WE MOVE FORWARD WITH HOPE?

Recognizing that we environmentalists, owing to the dourness of our cause, run the risk of pushing away our would-be supporters, it can be tempting to gloss over the most discouraging projections and to oversimplify the complexity of the solutions in an attempt to protect our audience (and ourselves) from despair. This choice has potentially counterproductive outcomes, including an undermining of trust in the messenger, a theft of the opportunity to face the challenge with full knowledge and a numbing balm that may prevent true progress.

For years, the medical field has been grappling with the question of the role of hope in confronting terminal diseases and when, if ever, it is appropriate for a doctor to deliberately cultivate false hope. Hope Theorist C. R. Snyder has found that doctors who deliberately lie to patients about the odds of their survival do so in an attempt to lessen patients' distress and to help them hold on to hope. Snyder's opinion, however, is that the patients usually discover the truth about the low odds and, when they do, the crucial doctor-patient trust is broken. Snyder suggests that doctors and patients face the severity of the disease 'head-on . . . and agree to make the very best fight possible' (Snyder 2002, 268). Medical doctor Jerome Groopman supports this honest approach, which recognizes that often things do not work out for the best. Groopman (2005) believes that, by being clear about the seriousness of the disease, doctors can help their patients develop a clear vision, which not only allows them to make good choices, but also triggers a set of positive chemical changes in the brain and body. Abraham Verghese, a professor of health sciences who worked with patients with early autoimmune deficiency syndrome, found some patients were able to use the illness to find meaning in their lives by hoping for more simple things. Verghese quotes Roxy Ventola, a performance artist who died of autoimmune deficiency syndrome, 'I know time is precious. So I don't waste it any more. Things that matter to the uninfected aren't important to me. . . . Now I'm free to be the person I always wanted to be' (Verghese 2004, 12).

These arguments for honesty echo the thoughts of theologian Jack Miles, cited yet dismissed by Orr in his essay 'Hope in Hard Times' (2004), who imagines that, as humanity sees its end near, we may create 'a new kind of religion and a new kind of art', making our final hour our 'finest'. If we environmentalists deceive our audience about the severity of our challenge, even if we do it with the best intentions, we are not only undermining our position as trusted messenger, but we are also denying society (and ourselves) the opportunity to develop a 'clear vision', to be deliberate in facing the future and to be the people we've always wanted to be.

Dishonesty in environmentalists' communications can take many forms, including a narrative of progress. The environmental movement has experienced many victories, including the creation of the Environmental Protection Agency and the passing of the Endangered Species Act, the Clean Air and Water Acts and the Wilderness Act. The widespread celebration of Earth Day, the ubiquity of recycling bins and the fact that U.S. rivers are no longer catching fire all contribute to the perception that the environment is being stewarded. While it is important to celebrate victories (and to recruit donations and members through shiny annual reports), we environmentalists might have given the public a false sense of security that, paradoxically, might convince them that things are being taken care of without them, thus robbing them of the motivation to become agents of change, which is the pathway to hope (Blohdorn 2002). In truth, our instincts tell us that few people will want to join a movement with a narrative of loss and destruction.

The question clearly arises, then, of how we can be honest about the severity of our situation without driving people to despair, denial or distancing. How do we clearly lay out the challenge, convince people to join us and keep them engaged for the long haul? How do we create an honest hope that does not rely on naïve optimism, utopian visions or overhyped solutions?

Once we have brought the issue of environmental despair out of the shadows and created a framework for understanding it as something other than a personal failing, and once we have honestly assessed our situation, we have multiple tools for creating hope. These include replacing the narrative of tragedy with one of comedy, embracing its associated humour and attitude of play; adopting the philosophy of meliorism, with its focus on incremental change through human effort; and providing venues for sharing of mutual experience.

David Orr explores the benefits of a narrative of comedy in his essay 'Hope in Hard Times' (2004). He argues that, when we frame our struggles through the narrative of tragedy and its basic metaphor of warfare, we take ourselves too seriously, failing to recognize our limitations, and we remain bent on 'vain attempts to reshape the world'. Framed as the tragic heroes, Orr argues, our fight against the ecological crisis is irrational and pathetic rather than noble. The comic narrative and its basic metaphor of a game, in contrast, allow us to recognize our limitations, laugh at our foibles and 'trim the sails of our ambition accordingly' (Orr 2004, 297). The comic view does not insist that we be in control; rather, it celebrates our land community and encourages us to learn more about the other members, both human and nonhuman, so that we may better fit in with them.

This comedic view has practical value. As mentioned earlier, it allows us to begin to work, even when we do not see a way to an absolute 'victory'. Furthermore, a comedic view opens the door to an attitude of play, which can boost creativity, promote divergent thinking and mental flexibility and increase intrinsic motivation. This can result in better performance on complex tasks, better problem solving and increased innovation (Shiu 2014, 195–96). The comedic view encourages us to find humour in our situation. Humour competitively inhibits negative thoughts, promotes a general sense of well-being and leads to positive emotions. These, in turn, lead to an increased ability to develop a strategy for action and to an increase in the perceived ability to overcome obstacles, two of the ingredients for true hope as defined by Hope Theory (Texas A&M University 2005). Humour also can help us deal with death and dying (Palmer 1993) and can give us the emotional distance we need to protect ourselves from emotional exhaustion (Ying 2007). Ironically, by taking ourselves less seriously, we can both perform better and stay engaged longer, which the severity of our ecological challenge demands that we do.

The comedic view also encourages us to celebrate and to enjoy life, even though we cannot bend it to our will. It opens a space for us to take time away from our work trying to save the land to, without guilt, simply play in it. This allows us to follow the advice of Ed Abbey, the self-described 'part-time crusader':

> Save the other half of yourselves and your lives for pleasure and adventure. It is not enough to fight for the land; it is even more important to enjoy it. While you can. While it's still here. So get out there and hunt and fish and mess around with your friends, ramble out yonder and explore the forests, climb the mountains, bag the peaks, run the rivers, breathe deep of that yet sweet and lucid air, sit quietly for a while and contemplate the precious stillness, the lovely, mysterious, and awesome space. (Noss, Cooperrider & Schlickeisen 1994, 338)

Not only will we be more emotionally, physically and spiritually healthy if we heed Abbey's advice, but we will also be more fun to be around. If one thing we must do as environmentalists is recruit others to our cause, being fun can't hurt.

The philosophy of meliorism is another tool for hopeful action. Meliorism advances the idea that the world is neither good nor bad in itself; it is only good or bad, and only gets better or worse, as a result of human intervention and action. Professor of Philosophy James W. Sheppard (2004) finds that meliorism's focus on incremental change allows his environmental ethics students to 'remain grounded and realistically cognizant of the depth and breadth of challenges' while at the same time keeping hope (221). Meliorism allows for a utopian ideal, but places it in the background, while foregrounding smaller and more realistic goals. It replaces the pessimism and defeatism of Eco-Marxism and the optimism and utopianism of Ecological Modernization with the 'ability to see the reward in small changes' on the path to a larger goal (Leuck 2007, 259).

One final suggestion for fostering hope is creating venues for sharing mutual experience. As discussed earlier, societal norms discourage us from discussing 'depressing' topics in social settings. Without appropriate venues for discussing these emotions, people who feel environmental despair can feel isolated. My experience giving presentations about hope and despair in environmentalism has convinced me that people, if given a safe opening, are eager to talk about their experiences with these emotions. Discussions about environmental despair can help create a sense of community among those in attendance, which can help maintain motivation and emotional well-being. We are, after all, social animals that draw strength and solace from communion with others.

## CONCLUSION

We environmentalists have a responsibility to acknowledge and address the tendency of our messaging to create negative emotions, including feelings of fear, guilt, disempowerment and despair. We need to remove any social stigma associated with environmental despair, create safe venues for people to explore these emotions and provide them with a theoretical framework for conceptualizing the causes of these negative feelings.

If we are to be effective, both in recruiting and retaining people for our cause and in developing solutions that are appropriate, we will need to conduct an honest and sober examination of our challenges and limitations. We cannot rely on naïve optimism, utopian fantasy or deceptive oversimplifications. We will need to develop an honest, yet hopeful, stance that provides us with a way to move forward with a 'spirit of thankfulness and celebration, without worrying about whether we will win or lose' (Orr 2004, 297).

1. See Craig Vetter, "He's Not Worthy: A Portrait of a Millionaire at the Crux," *Outside Magazine* (January 1997).

*Chapter Two*

# Hope and Pressure

Andrew Brei

Some years ago, during my dissertation defense, one of my committee members asked me a question that has stuck with me. It seemed innocent enough at the time—in fact, I was glad it was asked of me, amid all those other (seemingly) more difficult questions. But in the years since, I've realized the real depth of the question, and only now am I beginning to get a handle on it.

*Are you hopeful about the future regarding the natural environment?*

I was arguing for the conservation and protection of the natural world. I had asked my committee to read example after example of the ways in which humans damage and spoil the natural environment (to the detriment of our wellness, I added). I suppose I had painted a pretty bleak picture of the state of things. But when asked whether I was hopeful about the future, I said that I was. *Why not?* I thought. *Seems like the right attitude to have. You can't fault me for being optimistic about an uncertain future.*

Of course, now I see that the question was at least as difficult as any other I was asked that day. And in order to answer it I'm going to focus on one likely reason for my hopeful answer: social pressure. I'm going to describe the pressure that exists in American society toward positivity, optimism and hope. I'm then going to reveal the extent to which that pressure extends to attitudes about environmental crises. I will explain why the hope that results from social pressure is not good for environmentalism, and I will describe and defend an account of hope that does a good job of capturing what it means to have genuine, justified hope in the face of a grim ecological reality.

## POSITIVITY IN AMERICAN CULTURE

Back in 1959, Viktor Frankl pointed out that 'it is a characteristic of the American culture that, again and again, one is commanded and ordered to "be happy"' (Frankl 2006, 138). And he is as correct now as he was then. Optimism, happiness, positivity, hopefulness—whatever marker you care to use, the general idea is the same. We are pressured to look on the bright side of things. The pressure may not be as overt as, say, a pop song extolling the virtues of not worrying, but instead being happy. But the pressure exists all the same. It pervades American society, in the words of Barbara Ehrenreich, '[l]ike a perpetually flashing neon sign in the background, like an inescapable jingle' (Ehrenreich 2009, 47).[1] And Ehrenreich ought to know. In 2001, she was diagnosed with breast cancer and shortly thereafter became aware of the tremendous push toward positivity in the face of such a terrible disease. She scoured the hundreds of breast cancer–related websites for testimonials, noting how many of them involved humour, positivity and the refusal to use 'defeatist' terms like 'patient' or 'victim'. 'Instead', she points out, 'we get verbs: those who are in the midst of their treatments are described as "battling" or "fighting", sometimes intensified with "bravely" or "fiercely"' (Ehrenreich 2009, 26). And the language of breast cancer isn't the only place where positivity is on display. There is a seemingly endless array of purchasable items meant to inspire and uplift—from apparel and jewellery featuring the pink ribbon to candles, wind chimes, and teddy bears. In Ehrenreich's view, all of this contributes to what she calls 'the cheerfulness of breast cancer culture' (Ehrenreich 2009, 27).[2]

But those struggling with serious illness aren't the only ones who are urged to be hopeful and positive. Losing weight, according to several websites, is as much a matter of diet and exercise as it is about a positive attitude and visualization.[3] According to Donald Hensrud, MD, a blogger for the Mayo Clinic, having 'the right attitude and feeling like you're on an enjoyable and sustainable journey will get you well on your way [to losing weight]'.[4] Finding a mate can be challenging as well, but it is certainly not going to happen with a negative attitude. Being positive, says the publisher of *Online Dating Magazine*, is the key to finding the 'right' person.[5] Even those who are interested in having more money are encouraged to think positively. Hundreds of self-help books exist that explicitly link having a positive attitude with 'attracting' more money.[6] So whether you're fighting cancer, aiming for your target weight, searching for your soulmate or pursuing riches, the pressure to be positive and hopeful is undeniable.

In the corporate world, we can see the same sort of pressure. The 1980s and 1990s saw plenty of what is called 'downsizing', resulting in unemployment for around thirty million Americans (Ehrenreich 2009, 114). In that climate of uncertainty and downward mobility, most chief executive officers of large corporations were less interested in changing the realities for their employees than they were in changing how their employees *saw* corporate reality. The situation was bleak, but the message that came across was 'Work hard, distinguish yourself, and expect good things to happen'. Says Ehrenreich, '[t]his was the corporate world's great gift to its laid-off employees and the overworked survivors—positive thinking' (Ehrenreich 2009, 116). Indeed, perhaps the best indication of the pressure in America to be positive is the billion-dollar industry that has grown up in the last couple of decades around the promotion of positive thinking. *Motivation* is the commodity this industry peddles, and it is packaged in a plethora of ways. You can fill your home and/or office with posters, calendars, mugs and various other knick-knacks, all of them doing their part to inspire and motivate (with phrases like 'Dream Big' and 'Believe in Yourself'). You can also buy books, CDs and DVDs that will (it is claimed) pave the way to success with positive thinking. You can also attend any of a huge number of seminars, retreats and other interactive events designed to afford motivational speakers and 'life coaches' an opportunity to impart their message of hope and positivity—as well as sell their books, CDs and DVDs. A hopeful attitude and a positive outlook have become products to be marketed and sold to Americans.

Serious illness, the challenge of weight loss, the tribulations of dating, economic troubles—positivity and hopefulness are the prescribed attitude in all of these cases. The message our society is spreading is this: no matter how dreary a situation appears, the *wrong* thing to do is despair. Instead, think positively, envision success and be hopeful! Now, I don't doubt that there is something of value in this advice. For many people, personal experience will have demonstrated the connections between despair, depression and inaction. And plenty of studies have established a link between mental attitude and physical wellness. It is certainly difficult to accomplish goals when one's outlook is grim—and it is nicer, generally speaking, to be around those who are upbeat and hopeful. So there is something to be said for having a hopeful, positive attitude. But there is reason to be cautious here. In a recent work, Gabriele Oettingen describes how having positive fantasies involving some desired outcome can make it *less* likely that the outcome will be realized. In other words, when people allow themselves to dream (hope, fantasize) about something they want to happen—landing a new job, going on a date, running a marathon—they become less likely to do what they must in order to *make* that thing happen. Dreaming and hoping can placate a person to the point of inaction, leading to frustration and dissatisfaction.

The reason for this seems to have to do with the body's physiological response to positive fantasies. By allowing ourselves just a few minutes of daydreaming about some hoped-for situation, we can achieve the kind of relaxation usually associated with meditating, taking deep breaths or getting a massage. A measurable drop in systolic blood pressure accompanies positive fantasizing, the result of which is relaxation and calmness (Oettingen 2014, 42). So by allowing ourselves to indulge in such fantasies, we benefit psychologically and physiologically—but we do ourselves a disservice when it comes to goals that require taking action. It may seem counter-intuitive, but dreaming counter-productively robs us of the energy and motivation we need in order to bring about many of our personal goals! In light of this, Oettingen concludes that our hopes are more likely to be realized when we base them on reasonable assessments of past experiences. I will return to this idea below, as the idea of a 'reasonable assessment of past experiences' requires some clarification. For now, bear in mind that when it comes to achieving the outcomes we hope for, there appears to be a 'sweet spot' between paralyzing despair on one hand and pacifying fantasy on the other.

As I said above, Americans are compelled to be positive and hopeful, no matter how dreary a situation appears to be. Now, when I think of dreary situations, I tend to think of environmental degradation, habitat loss, climate change, and the like. I think back to the bleak picture of the natural environment I painted during my dissertation defence—and I think about how much bleaker things have gotten since then. It occurred to me recently to investigate the extent to which positivity and hope operate in environmental thought.[7] And what I found was both interesting and worrisome.

## HOPE AND ENVIRONMENTAL THINKING

In March 2014, the Intergovernmental Panel on Climate Change released a report that highlights many of the likely consequences of global warming, including flooding, droughts, heat waves and food shortages. The Intergovernmental Panel on Climate Change seems to have had two reasons for pointing out these unfortunate effects. First, it warned that governments around the world are ill prepared for dealing with these catastrophes and the upshots of them (including social and political instability). Second, it wanted to encourage people to be positive in the face of climate change. Chris Field (one of the report's authors) recognizes the connection between despair and inaction. Says Field,

> One of the things that made it so difficult for individuals and countries to be serious about climate is that the agenda is such a downer.
>
> If climate change is a total downer because everything looks so serious, and the only ways to cope effectively are to give up all good things in life, it's going to be really hard to take action. (Gosden 2014)

So Field suggests that policymakers be more positive as they deal with environmental challenges. He recommends that we focus not on the gloomy facts, but rather on the "really exciting opportunities" afforded to us by climate change. The opportunities he has in mind involve technological changes in the realms of energy, housing and transportation. Adjustments in these areas of our lives are key to reining in climate change, and if we allow ourselves to despair over global warming, then, says Field, 'we are not going to be able to attract the creative people who build the solutions'.

Sociologist Ulrich Beck displays a similarly positive attitude toward climate change. He suggests that the debate over the anthropogenic nature of climate change is a waste of time, and he recommends that we proceed *as if* humans were the primary causes of changes to the Earth's climate patterns. Why this Pascal-esque stance?[8] Beck believes that the benefits of carrying on as if the primary driver of climate change were human activity are very good indeed, far outweighing the drawbacks.

> Under conditions of uncertainty, even if we never know whether climate change is real, it is still pragmatically better to decide to accept that it is real. By taking this perspective we can bring about all kinds of social and political innovations and transformations which [*sic*] benefit the world. (Beck 2014)

Positive thinking has even found its way into academic ethics! Keith Douglass Warner and David DeCosse (of the Markkula Center for Applied Ethics at Santa Clara University) tell us:

> Virtue ethics challenge us to *move beyond our negative feelings* and focus on what kind of person we want to be, what kind of character will help us live out our commitments. This kind of *hope*, rooted in our habit of mind *and heart*, is precisely what we need to bring to situations where environmental injustices are being perpetrated. (Warner & DeCosse 2009: *my emphases*)

Two things stand out here. One is the fact that among writers on environmental matters, there are some who make a point of emphasizing the silver lining of ecological problems. These are the same people who would regard 'problems' as the wrong word to use here, who would prefer to use words like 'challenges' or 'opportunities' in order to show that motivation isn't upset. Speaking of motivation, the other thing to notice is the reliance on a connection between hope and motivation to act. Addressing environmental crises requires action, on both large and small scales. And actions are the products of motivation. Knowing that changing minds is the best way to change behaviour,[9] many environmental leaders (politicians, scientists and other experts) speak of optimism,opportunity and hope. They do this in order to catalyse and motivate, knowing that they run the risk of engendering despair if they merely lay out the grim realities of environmental problems. As I said above, I believe that there is value in this way of thinking. Being hopeful and positive can be beneficial. But I also believe that there are reasons to be cautious when it comes to estimating the importance of hope in our reactions to environmental issues. Two reasons, in fact, come to mind: one has to do with hope's power to *de*motivate, and the other has to do with the tension that exists between positive messages of hope and the ecological crises we face.

Can hope really have a detrimental effect on motivation and positive change? Douglas Main (2013) reports on a study that suggests it can. According to this study, 'when people have positive feelings toward climate change, such as hopefulness, they are more likely to avoid seeking information about it'. The research, conducted by Janet Yang at the State University of New York, Buffalo, points to a social norm involving information avoidance. If I keep getting hit with the message 'Be hopeful that climate change will not be so bad', then I am at once likely to be hopeful and *unlikely* to look any deeper into the matter. Positivity may derail positive action by undermining motivation with a false sense of security and optimism. Yang's conclusion is that instead of establishing associations between environmental crises and an attitude of hopefulness, it might be more effective (as far as bringing about positive changes goes) to link environmental issues to their alarming and uncomfortable consequences.

This is the same sort of phenomenon I described above relating to Oettingen's work on the psychology of positive thinking. Recall, Oettingen's research established a link between positive fantasizing (imagining a hoped-for outcome and the good that could come from having achieved it) and a reduction in the likelihood of achieving the object of that fantasy. A striking example from Oettingen involves the shortage of medical resources in Sierra Leone. In a study she conducted with Heather Barry Kappes, two groups of subjects were randomly created and presented with the facts of Sierra Leone's struggles. One group was asked to fantasize that the shortage of medical supplies had come to an end; the other group was not. The upshot is that the subjects in both groups were equally likely to donate $1 to the relief efforts—but the subjects who had imagined that the problem had been happily solved were *less* likely to donate $25 than were the subjects who had simply been given the facts. The results of this study are consistent with the many other studies Oettingen and her collaborators have conducted. Positively fantasizing about hoped-for outcomes makes people less energized and less likely to realize their hopes and dreams.

Clearly, this is important to bear in mind when we think about ways in which ecological crises might be addressed. Encouraging hope by way of fixating on (and fantasizing about) the end of environmental devastation can be counter-productive. If we spend too much time enjoying rosy thoughts of ecological harmony, then we may very well become placated by those thoughts to the point where we fail to recognize and perform the actions necessary for creating that harmony.

The other reason I prescribe caution when assessing the importance of hope has to do with the tension that exists between positive messages of hope and the facts associated with the ecological crises we face. When placed in a context of a society that endorses hope, information from scientists about the dire state of the natural world can cause a kind of cognitive dissonance. On one hand, we're told about how quickly and permanently things are getting worse. On the other hand, we're told to be hopeful that things will get better. This sends a mixed message that is wonderfully captured by one of the *Onion*'s headlines: '"It's Not Too Late To Reverse The Alarming Trend Of Climate Change", Scientists Who Know It's Too Late Announce' (*The Onion* 2014). Speaking of the same kind of mixed message, Michael Nelson puts the problem rather more concisely: 'We are bombarded with a contradiction'.[10]

Standard psychological responses to contradictions of this sort include coping mechanisms like *avoidance* and *masking*. When forced to reckon with conflicting beliefs or instructions, some choose to ignore and avoid part (or all) of the conundrum. Such would be the case if Sally learned that her favourite dry cleaner was polluting a nearby stream, but ignored that fact because of the top-notch service she received there. So too, I suggest, might one deny the overwhelming evidence that humans are a major factor in the warming of the Earth's climate merely because of what that would require by way of economic, social and lifestyle changes. When one's way of life is challenged by new information, it's often easier to dismiss the information than it is to change one's life.

Furthermore, the pressure to hope can cause one to mask one's true feelings concerning environmental crises. If you are like me, thinking about deforestation, biodiversity loss, pollution and other ecological calamities is likely to make you feel badly. But if you (like me) live in a society that encourages hope and positivity, then you are also likely to feel some pressure to suppress those bad feelings along with any expressions of gloom they might produce. A society that encourages hope and positivity also encourages masking negative emotions with the expression of positive ones. Thus, my sadness and frustration over ecological problems might be masked by my expressions of hope for a better future and optimism regarding people's willingness to make positive changes. This reminds me of the social norm—widely observed in America—according to which the standard answer to 'How are you?' is 'Fine', regardless of whether that happens to be the correct answer at the time.

This might not be such a bad thing, but for the possible negative consequences of masking. Some psychologists suggest that masking negative emotions can result in stress, resentment (over the need to mask), feelings of inauthenticity and mental exhaustion due to the 'emotional labor' required (De Gere 2008; Fischer & Ashkanasy 2000; Von Scheve 2012). And to the extent that these feelings are inconsistent with being highly motivated, they run counter to hope's role as a motivator of positive action in the face of ecological crises.

None of this is to say that all hoping is a waste of time! I am merely suggesting that we be cautious about why and how our hopes arise. If a person is hopeful simply because she feels that she "ought" to be (or because it is expected of her), then her hope lacks potency. The hope that results from social pressure is likely to be as false as the smile that results from a photographer saying, 'Cheese!' It is as Viktor Frankl said:

> If you want anyone to laugh you have to provide him with a reason, e.g., you have to tell him a joke. In no way is it possible to evoke real laughter by urging him, or having him urge himself, to laugh. (Frankl 2006, 138)

I am once again reminded of Barbara Ehrenreich's account of the demands placed on people with cancer to be hopeful and positive. In describing the downside of this pressure, she says:

> But rather than providing emotional sustenance, the sugar-coating of cancer can exact a dreadful cost. First, it requires the denial of understandable feelings of anger and fear, all of which must be buried under a cosmetic layer of cheer. This is a great convenience for health workers and even friends of the afflicted, who might prefer fake cheer to complaining, but it is not so easy on the afflicted. (Ehrenreich 2009, 41)

Ehrenreich knows better than I do what to say about hope's role in the context of diseases like cancer. What I know is that fake hope is not what we need in the context of environmental crisis. What we need is genuine hope. And as Frankl suggests, having genuine hope is going to depend on having *reasons* to be hopeful. I think this is akin to what Oettingen is suggesting when she says that our hopes are more likely to be realized when we base them on reasonable assessments of past experiences. I think we would do well to conceptualize hope—genuine hope—as an attitude that requires justification. And I think that among recent accounts of hope, there is one that is particularly well suited to environmental concerns: Ariel Meirav's external factor account. For the remainder of this chapter, I will describe this account, how it fares in light of some recent criticism and why it lends itself to analysing hope's role in environmental crises.

## THE STANDARD ACCOUNT AND THE EXTERNAL FACTOR ACCOUNT

According to the standard account, hope involves two components. When one hopes, one desires an outcome or state of affairs *and* one believes that it is in some sense probable that this outcome or state of affairs will occur.

Proponents of the standard account[11] disagree on exactly what this probability must amount to, but most agree that hope involves the desire for something that is neither impossible nor inevitable. In other words, one cannot properly be said to hope for something that cannot happen *or* that must happen. In between these extremes lies hope as well as despair. On the standard account, when one hopes, one desires something and believes that this something might occur. And when one despairs, one desires something but does not believe that this something might occur.

According to Ariel Meirav (2009), this account is inadequate. His objection to the standard account is based on the fact that it cannot distinguish between hope and despair in every case. There are situations, says Meirav, in which two people desire some outcome, assign the same probability to that outcome's occurring, but where one person hopes and the other fails to hope. Here is one of the examples Meirav offers to illustrate the problem:

> Suppose I buy a lottery ticket, and come home full of enthusiasm, showing the ticket to my wife and wanting to share with her my great hope in winning a sizeable monetary prize. But she is unconvinced, and her sceptical gaze expresses an amused indifference. Of course, she desires the extra income no less than I do. And we do not disagree on the probabilities we assign to winning. . . . And yet I am hopeful of winning, and she is not. (Meirav 2009, 223–24)

To borrow from another of Meirav's examples, we might state the present case like this: Meirav's wife will say, 'I grant you that it is *possible* to win, but the chances are one in a million!', while Meirav himself will say, 'I grant you that the chances of winning are one in a million, but it is *possible*!' 'The whole of the contrast between hoping and despairing', says Meirav, 'seems to be expressed in the subtle difference between the intonations of the word "possible" in these two sentences' (Meirav 2009, 223). Indeed, the example does a nice job of revealing the standard account's inability to delineate hope from despair in cases that are identical with respect to desire and probability. And if the standard account fails in this most essential of regards, then surely an alternative is needed.

In formulating an alternative, Meirav focuses on the nature of the desire that is involved in hoping. To understand that nature, suppose that I desire a sip of coffee from the mug that is sitting on my desk to my right. (Suppose too that I am aware that there is indeed coffee in the mug.) It would be very odd for me to say to myself, 'I hope I will have a sip of coffee from that mug.' The use of the word 'hope' seems out of place in that scenario. All I need to do in order to fulfil my desire is to reach over, grab the mug and bring it to my lips. I don't need to hope—I just need to act. In other words, I am the relevant causal factor in my desire's being fulfilled. Contrast this against cases in which I would not be the relevant causal factor: I hope the mail will be delivered soon, I hope the traffic will not be heavy for this morning's commute, I hope that my favourite band's next tour will stop through my hometown, and so on. In each of these cases, there is a desire for something that is beyond my control. Meirav calls this sort of desire *resignative desire*, and he says that '[w]hen I hope for a prospect, I desire it while viewing it as beyond the reach of my causal or epistemic powers' (Meirav 2009, 228). It is this sort of desire that is at work when one hopes, and it is this understanding of desire that allows Meirav's account to distinguish between hope and despair.

How so? Well, consider again Meirav's lottery example. The difference in attitude between Meirav and his wife—he is hopeful and she is not—cannot be explained by appealing to their desire or their probability assignments. As stipulated, these are the same for both. The difference in attitude is instead a result of different understandings of the external factors at work. Meirav and his wife may not share an understanding of exactly which external factors are relevant to the outcome in question. Meirav's wife may see this as a matter to be settled merely by the laws of physics governing the method of number selection (a numbered card being selected from a basket, a numbered ball being sucked into a tube, etc.). Meirav, on the other hand, may regard the laws of physics *as well as* some other factor as being relevant to the outcome (say, some cosmic force like Fate or God). Additionally, Meirav and his wife may differ with regard to their conception of how favourable or amenable the external factors are to the outcome in question. (Meirav uses the word 'good' to describe an external factor that is thought to be supportive of one's desires.) The laws of physics might reasonably be seen as 'indifferent' regarding the winner of a lottery, while Fate (or God or 'the Stars') might be seen as 'supportive' of a particular outcome (i.e., good). Note that on Meirav's account of hope, it doesn't matter whether Fate or some other possible factor *actually* has a hand in determining the desired outcome. What matters is whether a person regards something *as* an external factor and *how* that factor is understood in relation to one's desire for an outcome. Thus, certain instances of hope involve resignative desire, probability assignment, along with a positive assessment of the relevant external factors. In Meirav's words, '[i]f one views the external factor as good, then one hopes for the prospect. If one views it as not good, then one despairs of it' (Meirav 2009, 230).

Before I explain why I think this is a fine way of understanding hope when it comes to environmental matters, I want to address a criticism of the external factor account. In her recent book, Adrienne Martin (2014) says this about Meirav's account:

> Note one tricky thing about the External Factor analysis: believing an external factor operates on one's behalf can raise the probability one assigns to the

> hoped-for outcome. However, if that is what is going on in these cases, the orthodox definition [i.e., standard account] is a sufficient analysis of them. (Martin 2014, 19)

According to Martin, seeing an external factor as good can cause an increase to one's probability assessment. And if that is the case, then the external factor account collapses into the standard account. Meirav, it seems, recognizes this possibility when he mentions that *some* combinations of desire and probability assignment are enough by themselves to determine whether one hopes or despairs (Meirav 2009, 231). Yet he denies that this sort of direct causal relationship *always* holds.

Imagine that I am in the hospital, in pain and hoping the nurse will administer a painkiller soon. To say this (according to Meirav's account) is to say that I want her to do so; I assign a probability to her doing so of, say, around 30 per cent, and I regard her as good (meaning that I believe she wants to help me and is capable of helping me). Says Meirav, '[i]t might be thought that since I assume that the nurse is capable of coming soon to administer the drug, and both wants to benefit me and knows how to do so (even if imperfectly), I must assign her coming soon a *high* probability. This is a mistake, however' (Meirav 2009, 232). After all, there are plenty of reasons why the nurse may not come by soon to ease my pain. She may not know of my pain; she may know of it, but believe that my interests are better served by waiting; or she may decide that somebody else's interests take priority over mine. Any of these would make the likelihood of the nurse stopping by soon quite low, but none would change my belief that she is good. Even if the unpredictable complexity of a hospital reduces the probability of my desired outcome, my hope remains because of the way I regard the relevant external factor. Thus, the relations between probability assignments, assessments of external factors and hope cannot be as straightforward as Martin claims.

Still, Martin has important things to say about the external factor account of hope, and I cannot do justice to her critique here. Let me simply point out that she agrees with Meirav that hope is an attitude that requires justification in order to be genuine. It involves understanding the reasons and causes relating to the outcomes we desire. And it amounts to a complex of attention, thought and positive feelings. I remain convinced that Meirav's account is a good guide for estimating hope's significance in the realm of environmental thought. The primary reason for this is the fact that so many environmental problems result from multiple, intertwined factors. Often, these factors are external to those of us who concern ourselves with the natural environment, and so we must resign ourselves to the fact that the causes of *and solutions to* environmental problems are larger than any individual. Does this mean that making substantive changes on a grand scale is next to impossible? Certainly not. It does mean, however, that having hope in the face of environmental crises will depend on one's assessment of the relevant external factors. In turn, that will require at least some investigation of and reflection on the factors in play and the degree to which they are supportive of the desired outcome.[12] For example, if industrial oil leaks are the chief threat to wildlife stability in the Gulf of Mexico, then one would have to have evidence of a willingness to take corrective action on the parts of oil companies in the Gulf before one could have reasons to be genuinely hopeful that the wildlife there will thrive. In this way, hope is not concocted as a reaction to pressure. It ensues from having a certain kind of understanding of the drivers of (and potential answers to) the environmental ills we want to correct.

## CONCLUSION

I have shown that there is pressure in American society to be hopeful and positive. I have also demonstrated how this pressure extends into attitudes about the environmental conditions that threaten the integrity, stability and beauty[13] of this planet. Based on these realities, I recommended caution when estimating the importance of hope in our reaction to these crises. Hope that results from social pressure is too often empty and impotent. We run the risk of derailing efforts to make positive changes when we succumb to external pressure to feel a certain way. Far better, I claimed, to regard hope as an attitude that requires justification by means of a positive assessment of the external factors involved. Looking at hope in this way, we shift our focus away from the outcomes we desire, toward the steps that will be involved in achieving those outcomes. This is what I regard as part of what genuine hope requires.

I now find myself in a position to answer the question I shared at the beginning of this chapter. Am I hopeful about the future regarding the natural environment? Well, I certainly desire an end to the destructive, short-sighted, unsustainable activities that are threatening the future. And I certainly believe that with a grand, concerted effort, such a thing is possible. But I also have the sense that the obstacles before us—ignorance, greed, irresponsibility, recalcitrance, indifference, fear, custom—are vast and complicated. And I neither presume to have a full understanding of the external factors involved, nor a clear sense of whether they are (in Meirav's sense) good. What I do know is that there seem to be more champions of environmental causes than ever, and my impression is that this trend is continuing upward. All that said, I must admit that my hope is not strong . . . but at least it's genuine.

# NOTES

1. I should note that Ehrenreich does not equate optimism with hope. I don't think they are exactly the same either, though I will treat them here—along with happiness and positivity—as members of the same family.

2. 'I didn't mind dying, but the idea that I should do so while clutching a teddy [bear] and with a sweet little smile on my face—well, no amount of philosophy had prepared me for that' (Ehrenreich 2006, 17).

3. See, for example, http://greatist.com/health/losing-weight-holistic-techniques.

4. http://www.mayoclinic.org/healthy-living/weight-loss/expert-blog/positive-attitude/bgp-20056443.

5. I wasn't aware of this magazine either. http://www.onlinedatingmagazine.com/self-improvement/beingpositive.html.

6. Napoleon Hill's *Think and Grow Rich* is the most conspicuous example, over seventy million copies of which have been sold worldwide.

7. By 'environmental thought', I mean attitudes as expressed in academic environmental literature, popular media (including news) as well as in environmental policy. I don't imagine that I've got a clear grasp of how hopeful (or not) 'everyday people' in America are as a whole.

8. I hope this reference is neither misplaced nor too opaque. Recall, Pascal advanced an argument for God's existence based on the values of the possible outcomes resulting from God's existing (or not) and one's believing that God exists (or not). See section 233 of Pascal's *Pensées* for the argument.

9. I realize that this is a rather broad statement. To adequately support it would sidetrack the train of thought we are on now. I'll simply point to something Aldo Leopold said in this regard (as I am wont to do): 'No important change in ethics was ever accomplished without an internal change in our intellectual emphasis, loyalties, affections, and convictions' (Leopold 1989, 209–10). In other words, change follows actions and actions follow mindsets.

10. See chapter 9 of this book.

11. Meirav mentions several proponents of the standard account of hope on page 217 of 'The Nature of Hope' (2009).

12. As I noted earlier, what matters on Meirav's account is whether someone considers an external factor to be relevant, regardless of whether it actually is. This means that I could take Poseidon's will as a relevant factor—and if I saw him as supportive of my desires, then I might be hopeful that rising sea levels will cease to threaten indigenous island cultures. This may seem like an odd consequence of Meirav's account, but it is no more peculiar to Meirav's account than it is to any other—or indeed, to the social pressure to be hopeful.

13. To borrow once more from Aldo Leopold.

*Part II*

# Hope and Moral Judgements

*Chapter Three*

# Playing a Requiem on the *Titanic*

## *The Virtue of Hope in the Age of Ecological Calamity*

Andrew Fiala

There is nobility and virtue in struggling and remaining hopeful, even in the face of calamity. Hope is not always efficacious. But hopefulness is a virtue—a habit and disposition that is a constitutive element the moral life. Nancy E. Snow has defined hopefulness (following Shade 2001 and others) as a dynamic orientation toward the future, which generally expects good outcomes. According to Snow, the attitude, disposition or virtue of hopefulness is 'the sine qua non of human life and agency' (Snow 2013, 153–54). Hopefulness may have a payoff in terms of good consequences. But hope is also an end in itself. It is useful and good to remain hopeful—even in the face of catastrophe.

A haunting metaphor of heroic hopefulness amid disaster comes to us from the sinking of the *Titanic*. As the great ship sank, the ship's orchestra played 'Nearer My God to Thee'. While the true motives of the musicians drowned with them in the Atlantic, one imagines that they stoically played on out of a sense of obligation to the escaping passengers and as a way of celebrating their own profession. The musicians performed their own requiem, comforting others, giving hope to survivors and remaining true to the virtues of their craft. This poignant image reminds us that even in a hopeless state, we can retain our dignity, maintain moral equilibrium and find small ways to make the world better. In the face of catastrophe, it may be reasonable to give up idealistic hope for a miraculous rescue. But hopefulness and associated virtues are essential components of finding a moral path through the teeth of calamity.

## OUR HOPELESS PREDICAMENT

We have reached important tipping points in recent years. Global levels of carbon dioxide have increased to unprecedented levels. Glaciers are melting. Species are going extinct. Land is deforested and desertified. Groundwater is being depleted. The oceans are becoming acidic. And so on. Despite the evidence of anthropogenic ecological calamity, it is not clear that humanity is willing to take the necessary actions required to roll back these radical changes to the ecosystem. Indeed, at this point, some of these changes will persist for decades and centuries. And yet the development of industrialized global civilization relentlessly continues.

The problem appears to be systematic. The system of global industrial capitalism appears to be incompatible with sustainable environmentalism. Paul Ehrlich has been warning for decades of a population bomb and the dangers of unsustainable development. He thinks that a collapse of civilization is likely if we do not change our ways. Erhlich writes (along with co-author Anne B. Ehrlich):

> Serious global environmental problems can only be solved and a collapse avoided with an unprecedented level of international cooperation. Regardless of one's estimate of civilization's potential longevity, the time to start restructuring the international system is right now. If people do not do that, nature will restructure civilization for us. (Ehrlich & Ehrlich 2013, 5)

Ehrlich's warnings have been reiterated by a number of scholars. Clive Hamilton explains in his dire book *Requiem for a Species*:

> Climate scientists themselves now evince a mood of barely suppressed panic. No one is willing to say publicly what the climate science is telling us: that we can no longer prevent global warming that will this century bring about a radically transformed world that is much more hostile to the survival and flourishing of life. . . . This is no longer an expectation of what might happen if we do not act soon; this will happen, even if the most optimistic assessment of how the world might respond to the climate disruption is validated. (Hamilton 2010, xi–x)

James Lovelock—famous for his 'Gaia Hypothesis'—appears to agree. Lovelock thinks we will be unable to fix the environmental problem and that we are nearing the end of an epoch, which will be brought to a final conclusion as Gaia strikes back (Lovelock 2006, 2008). Some experts are now calling the new Earth ecosystem or epoch 'the Anthropocene', 'Eaarth' or 'Earth 2.0', and speaking of 'the end of nature' (Cruzten 2002; McKibben 2006, 2010). Perhaps it is time to embrace the new world and give up hope of returning to Earth as it was before the Anthropocene. Paul Crutzen suggested, in the article in which he named 'the Anthropocene', that we can't go back. Instead, Crutzen suggests we have to learn to 'optimize' changes, perhaps by developing massive geo-engineering projects. When we give up hope for a return to nature, the next step is to take up the burden of being a 'caretaker' or 'custodian' of an Earth that has been recreated in the interest of humanity. The idea of taking up the burden of being a caretaker or custodian has led Bill McKibben, for example, to lament the demise of nature and the hubris of humanity: 'This intended rallying cry depresses me more deeply than I can say. That is our destiny? To be "caretakers" of a managed world, "custodians" of all life? For that job security we will trade the mystery of the natural world, the pungent mystery of our own lives and of a world bursting with exuberant creation?' (McKibben 2006, 182). McKibben points toward a kind of despair and dispirited acquiescence: 'If nature were about to end, we might muster endless energy to stave it off; but if nature has already ended, what are we fighting for?' (McKibben, 2006, 178).

McKibben begins his book *Eaarth* by admitting that the hopes of the environmental movement have routinely been dashed. He counsels 'maturity' about what we may reasonably hope for.

> Hope has to be real. It can't be a hope that the scientists will turn out wrong, or that President Barack Obama can somehow fix everything. . . . Maturity is not the opposite of hope; it's what makes hope possible. (McKibben 2010, xiv)

But might maturity now include acceptance of the fact that we cannot fix the problems we have created? Perhaps a mature hope requires us to give up hope for a soft landing. Instead, we may have to acknowledge a greatly reduced idea of what we may reasonably hope for.

Answers to the question of what we may reasonably hope for depend largely upon the *content* of hope. If we long for a return to the lush natural world of twenty thousand years ago, when large predators ruled the Earth and rivers ran freely through magnificent forests, that is a hopeless longing. But if we desire a world in which human beings continue to thrive in large numbers, that hope is possible to imagine. The current situation presents us with the need to revise our understanding of what we may—and what we ought to—hope for with regard to environmental issues. I suggest the following:

1. It is reasonable to hope that human beings can find solutions to some small-scale environmental problems. There have been some environmental successes—in the work of Aldo Leopold, Rachel Carsen, John Muir and others (both individuals and groups). Environmental movements have resulted in concrete improvements in some cases. We can each contribute in small ways, as the musicans on the *Titanic* did. Perhaps we cannot prevent the ship from sinking. But we can help

others survive—providing hope and serving as models of good stewardship.

2. But we ought to give up hope for a return to anything like pristine nature. We have crossed too many thresholds and tipping points for us to hope for a rescue or return. Human beings will simply not relinquish their place as the dominant species in the Anthropocene. This means that the dreams of Deep Ecology ought to be viewed with caution. There is no hope for Deep Ecology in a world of over seven billion human beings. And there is risk of misanthropic outcomes, if we were to remain wedded to unrealistic hope for a return to pre-industrial civilization.
3. Thus, we ought to direct our hope in an anthropocentric direction to include concrete actions taken on behalf of human beings—to ease the transition to the Anthropocene with an eye on global environmental justice and intergenerational justice. This anthropocentric concern does not go as deep as some would like. But our concern ought to be to find ways to mitigate environmental harms. This does not mean we should entirely jettison the idealism of nonanthropocentric ecology—or view it as metaphysical nonsense. Perhaps some day our values will radically change. But in the middle of disaster, our efforts should be focused on mitigation and maintaining respect for human dignity in the Anthropocene.

## SINGING IN THE MORAL STORM

Stephen Gardiner has described the present situation as 'a perfect moral storm' in his discussion of the moral problem posed by climate change (Gardiner 2011). One part of Gardiner's assessment focuses on corruption—where we rationalize bad behaviour and excuse it, even though we know it is bad. Gardiner also outlines problems such as the intergenerational prisoner's dilemma and the basic difficulty of thinking globally. Given these problems, Gardiner suggests that climate change may be a problem that we are morally unable to solve.

A significant problem, however, is that if Gardiner is right, then there is no reason to hope for a solution—and a vicious circle results. Hopelessness undermines positive action and moral responsibility. If we truly are in the midst of a perfect moral storm and if other assessments about the facts of climate change (and other ecological calamities) are true—that we have reached certain tipping points, which political and social power have been unwilling or unable to prevent—then it may be rational to give up hope and give up on meaningful moral action. Bleak assessments may actually cause further moral failure. As we give up hope for a solution, we may shirk moral responsibility—thinking that responsible action is useless. The challenge is inspiring responsible action, while also admitting to a realistic assessment of the ecological problems we must confront.

The worse things look, the more absurd hope becomes. I argued in a previous essay (Fiala 2010) that in a severe crisis it may be rational to give up hope and retreat to short-term self-interest. In that essay, I use prisoner's dilemma–type arguments to show that in the face of intractable eco-crises, it can appear to be rational to simply give up trying to reduce consumption—that there is some practical wisdom in 'fiddling while Rome burns'. Rational self-interest maximizers may argue as follows: 'If the crisis cannot be averted and everyone else is consuming without restraint, then what's the point in conserving . . . ?' However, morality requires that we not take the path of short-term self-interest. Indeed, prisoner's dilemma arguments point to the problem of egoism: short-term self-interested rationality leads to worse outcomes. Thus we ought to hope that moral action can make a difference in producing good outcomes—or else we will end up with rational hopelessness that is self-affirming (and produces worse outcomes). It is possible that hope can be grounded in a sort of long-term and globally oriented consequentialism. Hope is useful. But a stronger argument grounds hope in a theory of duties, virtues and responsibilities. We have a responsibility to behave morally, even in hopeless circumstances. One ought to pursue eco-friendly behaviors, lifestyles and political action—even though these are irrational as understood from the perspective of short-term self-interest.

Let's contrast the myth of Nero fiddling while Rome burns with the image of *Titanic*'s orchestra playing while the White Star Line vessel sank. In the face of inevitable catastrophe, an egoist (such as we imagine Nero to be) reasons that he might as well enjoy himself before catastrophe engulfs him. If there really is an impending catastrophe, we might adopt the motto, 'eat, drink, and be merry for tomorrow we shall die'. The insight of this proposition becomes more obvious to the egoist when the inevitability of the catastrophe becomes clearer. And indeed, we might imagine a modern-day Nero, with a massive carbon footprint, feeling a malicious sort of superiority as he watches others skrimping and saving in a hopeless effort to stave off calamity. One way of understanding this is based upon a consideration of utility maximization, in hopeless conditions. It makes little sense to sacrifice short-term happiness in the name of hope for improbable outcomes. Nero is a rational egoist.

The members of *Titanic*'s orchestra are, however, not egoists of this sort. They also recognize a hopeless situation. But they choose to complete their professional duties. They choose, through their behaviors and encouraging song, to give hope to others. And they perish while fulfilling their calling. These are virtuous individuals, motivated by a set of ideals about duty and dignity. Although they are unable to prevent the sinking of the ship, they are able to facilitate a humane and optimal (given the conditions) transition to the next phase. Indeed, survivors reported being inspired and thankful for their noble effort.

Here is the problem: the worse things are, the more hope is needed; however, as things become worse, hope appears to become irrational. At some point, it is rational to give up hope. And yet, morality of a certain sort—that which is based upon duty and virtue—appears to require hopefulness, despite its irrationality. It is possible to imagine grounding hope in consequentialist concerns. We might praise *Titanic*'s musicians for facilitating the orderly evacuation of the ship. Some form of hopefulness prevents panic: it is clearly important for producing good outcomes and for solving some collective action problems. But Kantian ethics and virtue ethics provide a clearer account of the value of hopefulness.

Kant indicates that morality requires hope and that hope is connected to positive, constructive and morally appropriate behavior. Kant holds that faith in God, belief in freedom and hope for immortal life are central postulates of the moral life. Kant also affirmed a teleological account of history: Kant's hope for progress was grounded in his idea that the age of Enlightenment was dawning and the world was becoming a better place (see Fiala 2002). One need not go so far as to affirm all of Kant's postulates to reach the conclusion that some form of hope is basic to moral action, duty and responsibilty. From this standpoint, there is a moral imperative to believe ecologically responsible behaviour will make a difference. Some versions of deontological duty may even claim that duty is required even in the face of hopelessness.

Virtue ethics also teaches that hopefulness is a virtue. This is clear in Catholic teaching, which follows Aquinas in understanding hope as a theological virtue (which we will discuss again below). Aristotle is less explicit about hope. But we can find ideas about hope in the context of Aristotle's discussions of courage, confidence and fear. For Aristotle, a sort of hopefulness is important for moral behaviour and improvement (see Gravlee 2000). It is a virtue of the great souled person (*megalopsyche*) to believe that his or her deeds matter. Of course, the Stoics correct this somewhat by denying that hopefulness is as important as gritty endurance and perseverance. Stoicism emphasizes seeing the world as it is and not as we want it to be. This fatalistic affirmation nonetheless requires us to hope that through our own efforts, we can be virtuous—despite the adversity of circumstances. But it is important to note that stoic hope is directed inwardly—as a hopeful attitude toward one's own virtue. Stoic resolve may help explain what happens in the case of the *Titanic*'s musicians: they were not naïve about the circumstances—and yet they stoically played on.

On the other hand, an empirically minded consquentialist may ask why we ought to think that morality requires the sorts of leaps of faith about hope and an ultimate payoff for moral action that Kant defends. If morality does not produce obviously good conseqences—but requires hope in infinitely deferred gratification of the Kantian sort—then morality asks too much of those of us who live in the real world. From a consequentialist standpoint, what matters is maximizing good outcomes. This includes a balance of both short-term and long-term goods. Consequentialism is grounded in empirical claims about present reality. For the consequentialist, morality is less about hope than about plausible predictions; it is less about what is required by duty than about what we can reasonably expect will produce beneficial outcomes. And with regard to virtue, a consequentialist would not care much about stoic interiority—since the consequentialist is more concerned with external consequences. If stoic musicians help others to escape disaster, that is good. If not, then their performance is futile.

Notice that hope on the Kantian account is understood in transcendental terms: there may be no good empirical reason to believe that such action will pay off, but we ought to hope it does nonetheless, as a transcendental condition for the possibility of moral action. The Kantian duty to take positive action and to remain hopeful transcends what is merely understood in terms of rational self-interest and what is provable within the empirical world. Kantian morality emphasizes responsibility that points beyond a merely empirical account of what works or does not work. Understanding the normative and transcendental feature of moral responsibility helps open the door toward a larger point of view that fits well within a broad and deep form of environmental ethics. Deep ecologists, biocentrists and nonanthropocentrists ask us to consider a larger perspective than mere short-term rational, human interest. This may be understood in terms of obligations to future generations or in terms of the value of nonhuman beings—animals, species and ecosystems.

Understanding the moral imperative of hope in Kantian transcendental terms thus opens the door to further understanding of the normative basis of ecological concern for future generations and nonhumans. The point here is that a moral responsibility to future generations is based upon a kind of hope—a projection into an uncertain future that transcends the present. It is easy to be hopeless, to ignore future generations and to focus on narrow empirical concerns. However, morality requires that we take action now with the belief that our present actions can make the future better—even for those who are not yet born—and morality requires that we project our concerns beyond what is warranted by empirical generalization, even when such projections appear irrational from the standpoint of empirically based consequentialism.

However, the consequentialist response to the deontological defense of hope is cutting. What if the crisis or catastrophe really is as bad as some predictions make it out to be? What if hope really is irrational? From a consequentialist standpoint, it is better simply to accommodate ourselves to the new reality of the Anthropocene, better to stop worrying about what we've lost, better to stop beating ourselves up about our moral failures and better to find a way to live well within the parameters of the new empirical reality. The consequentialist is a meliorist—focued on finding ways to maximize good outcomes within present constraints. Meliorism is about improving the situation incrementally: it is about making things better within current parameters. But meliorism gives up on idealistic notions of duty and abstract ideals about responsibility. The deontological imperative of hope relies on abstract postulates that point in the direction of metaphysical wishful thinking. For a consequentialist, all of this idealism gets in the way of taking practical action to respond to the needs of the present, changing situation.

Here we might bring in the concerns of virtue ethics. Hopefulness is a feature of our character, habits and dispositions. It is derived from our roles and duties. And it is linked to all of the other virtues. As human beings, we have an obligation to ourselves and to others to behave nobly, courageously and hopefully, even in the teeth of disaster. Considerations of consequences matter. But so does our moral evaluation of who we are and what sort of charater we exhibit. Greek virtue ethics—of either the Aristotelian or Stoic variety—does not require Kantian faith in transcendental reconciliation. Instead, it is grounded in more limited concerns about our roles and functions as rational, social beings. From this standpoint, a hopeful demeanor is praiseworthy—as a service to the community and a testament to one's own virtue.

## METAPHYSICAL HOPE?

Metaphysical and religious speculation may easily enter into discussions of ecological hope. Perhaps we ought to hope that God does not abandon us—as Pope Francis has recently explained in his Encyclical *Laudato Si*. Francis clearly states that Christian faith means that we are 'not adrift in the midst of hopeless chaos' (Francis 2015, para. 65). Kant might agree. But if we adopt a humanistic perspective, such metaphysical speculation is off the table. Moreover, from a humanistic perspective that takes seriously the findings of natural science, there is no assurance of good outcomes. The facts of geological history remind us that past civilizations have collapsed and species go extinct. Perhaps metaphysical hope can reassure us that everything will work out all right in the end. But metaphysical hope can mislead. And it may be more honest to admit that natural science does not give us any assurance that our dreams will come true.

Kantian morality and Kant's teleological account of history move beyond what science allows, however. One way of putting this is to understand that hope and progress require us to behave *as if* morality was rewarded or *as if* history was progressive. Kant makes this clear in his reflections on the nature of the moral law. He concludes the *Groundwork* by stating that in the Kingdom of Ends we would behave *as if* the moral law had the power of natural law (*als ob sie Gesetze der Natur wären*) (Kant 1870, 199). Hope is clearly a sort of 'as if' proposition. Phillip Pettit explains that hope (and its opposite, precaution) are both based upon suppositions about how the world may (or may not) turn out, which guide us to behave 'as if' the world was one way or another. He states, 'Hope would take a parallel form to precaution, then, if there were any reason in such a case why the agent should be moved to act as if that desired prospect were going to obtain or as if there were a good chance that it was going to obtain' (Pettit 2004, 157). Pettit goes on to explain that one good reason to act as if a good outcome were going to be obtained is that hopeful action may be more efficacious. 'This is the danger that the confidence level will be so low that the agent loses heart and ceases to exercise agency effectively. If the agent assigns a relatively low probability to the desired prospect, then that may cause him or her to make no effort to bring it about, thereby ensuring that he or she certainly does not bring it about' (Pettit 2004, 157). In the context of Kantian morality, we behave as if morality will be rewarded, as if we have free will and as if we think that moral action will produce the sorts of benefits we expect. With regard to history, we ought to behave as if progress is possible. The moral imperative to have hope can be understood from this standpoint as a useful heuristic: hopelessness enervates and devitalizes moral agents, while hope inspires us to fulfil the obligations of morality.

Pope Francis's recent ecological encyclical grounds hope in theology that appears to share much with the Kantian account. But for Francis, the assurance of theology is not a merely 'as if' proposition. It is a fundamental postulate of Christian faith: the joy of Christian hope is a metaphysical first principle. Francis concludes:

> May our struggles and our concern for this planet never take away the joy of our hope. God, who calls us to generous commitment and to give him our all, offers us the light and the strength needed to continue on our way. In the heart of this world, the Lord of life, who loves us so much, is always present. He does not abandon us, he does not leave us alone, for he has united himself definitively to our earth, and his love constantly impels us to find new ways forward. (Francis 2015, para. 244–45)

Christian hope can be traced to Paul's first letter to the Cornithians, where hope is arrayed in conjunction with faith and love. In *Summa Theologica* (Second Part of the Second Part, Question 17), Aquinas clarifies that hope is a theological virtue—it relies upon God, who provides the means to achieve the hoped-for end (which according to Aquinas is ultimately eternal happiness). Theological hope occurs in relation to the concrete metaphysical anchor of God, who is the last end and the first efficient cause of all things.

heological hope is thus oriented toward an eschatological account of the beginning and end of time, as Jürgen Moltmann has shown in *Theology of Hope* (Moltmann; see Fiala 2000). Theological hope occurs within an account of creation that views history as progressive and linear. Theological hope involves reflection upon what was lost in the beginning of history and expectation about what can be attained in its end. From within a theological and eschatological understanding of hope, each moment within history is a moment of crisis—a decision point at which one can either choose to progress toward our proper good and divinely ordained end or continue in sin and denial of human teleology and cosmic eschatology. It is not surprising, given the structure of the Christian understanding of hope, that Francis's encyclical connects the environmental crisis with a larger spiritual crisis. Francis explains:

> If the present ecological crisis is one small sign of the ethical, cultural and spiritual crisis of modernity, we cannot presume to heal our relationship with nature and the environment without healing all fundamental human relationships. (Francis 2015, para. 119)

The solution from this point of view is clear: ecology involves spiritual renewal—including ethical, economic, social and political justice. From this point of view, bleak environmental news creates an ordeal of faith. The heart of Christian eschatology contains an account sin, suffering, death and redemption. Thus the bleaker things become, the more hope is required as part of spiritual transformation. The Kantian account is perhaps less demanding—since Kant offers some reasons for historical hope in his account of history. But the Kantian account also points beyond the empirical world toward postulates of a rational, moral faith.

There is something admirable in these accounts of metaphysical hope. The Western tradition is infused with this idea. It is not insignificant, for example, that the musicians on the *Titanic* played 'Nearer my God to Thee'. But this deep metaphysical hope raises the stakes and creates a dialectical danger. As the ecological crisis continues, Kantian and theological hope become a test of the spirit. One wonders whether Kant would be as hopeful about progress, if he were alive today and were able to witness our inability to respond rationally to the ecological problem. The dialectical difficulty of metaphysical hope is that if we ever lose faith we may sink into deeper despair. If only God can save us—and God does not appear to be saving us—then what do we do? To defer our hopes to the afterlife may inspire a Christian to strive harder to be ethical. But geological calamity can easily burst that bubble. Metaphysical hope—if it is understood as asking us to 'hope against hope'—makes it psychologically difficult to act and live well in the midst of a mounting crisis.

## PRAGMATIC MELIORISM AND RADICAL HOPE

Instead of metaphysical hope, a less stringent sort of *psychological* hope may be all that we need. Hope is the expectation of good outcomes. Nonmetaphysical hope hinges on a prediction of probability; justifiable and realistic hope is grounded in previous experience and background knowledge. Basic *psychological optimism* or what we might call *pragmatic meliorism* is the idea that acceptable outcomes may be created, if we work a bit harder, if we have faith in the future and if fortune smiles upon us. Hope is pragmatically and psychologically beneficial—as so-called positive psychologists and advocates of the 'growth mindset' teach us (see Seligman 2006 or Dweck 2006). But a more radical, metaphysical sort of hope expects good outcomes *despite* the evidence. This sort of hope is often connected with *metaphysical optimism* and a kind of eschatological faith, as we saw above.

A different idea of 'radical hope' has been explained by Jonathan Lear, who connects it to ideas found in Kant and Kierkegaard. Lear explains that radical hope 'is directed toward a future goodness that transcends the current ability to understand what it is'. He continues, 'Radical hope anticipates a good for which those who have the hope as yet lack the appropriate concepts with which to understand it' (Lear 2006, 103). Lear explains that radical hope is connected to courage and imagination. Lear's analysis has inspired several other authors (Kretz 2014, Thompson 2010 and Williston 2012, for example) to delve into the idea of radical hope and its connection with moral action in the face of ecological crisis. Behind much of this is an acknowledgement of the need for readjusted expectations in light of radical ecological and cultural change. As Thompson explains, we will have to readjust our understanding of what counts as environmentally good in light of the new situation. 'A new form of an inherent goodness in nature and a new conception of living well are relevant objects of radical hope for the environmentalist' (Thompson 2010, 51). This idea of radical hope is that it is a kind of courage and hopefulness that remains engaged and responds creatively to radically changing circumstances.

This is a melioristic idea. We ought to adjust our hopes in a way that reflects realistic expectations about the world and about human agency. If we hope less, we can avoid hopelessness, by acting more and readjusting our expectations. Someone who is wedded to an eschatological ideal—or the ideal of a return to pristine nature—may see this as a less than optimal resolution to the challenge of hope. But Lear and others imply that we ought to readjust our ideals in light of changing circumstances. The idea of meliorism has deep roots in the American philosophical tradition—in the writings of pragmatists such as James and Dewey. Dewey, for example, defined it as follows: 'meliorism is the belief that the specific conditions which exist at one moment, be they comparatively bad or comparatively good, in any event may be bettered' (Dewey 1920, 178). The melioristic spirit can be found in James' essay 'Is Life Worth Living?': attitude matters when we go to accomplish a task. If you believe you are able to do something, you will be better able to do it: 'believe that life is worth living and your belief will help create the fact' (James 1956, 67). This motto appears to be required of moral action: we must believe that moral action can make a difference—and belief in the value of moral action will help us to act more ethically. As Kretz has explained, hope is a self-fulfilling prophecy that is efficaciously enabling (Kretz 2014, 926). She concludes: 'Hope is desirable for pragmatic reasons in that it can be self-perpetuating' (Kretz 2014, 942).

But not any hope will do. One significant concern is whether metaphysical hope and metaphysical optimism contribute to the crisis. 'Don't worry', a metaphysical optimist might say, 'Things will work out for the best'. Climate change deniers offer this sort of response when they assert, for example, that God would not permit us to destroy the Earth. In 2009, for example, John Shimkus (a Republican representative from Illinois) asserted in a Congressional hearing that according to the Bible, especially the story of Noah and the Flood, the Earth will not be destroyed in a flood: 'The earth will end only when God declares it's time to be over. Man will not destroy the earth. This earth will not be destroyed by a flood'.[1] Similar ideas have been offered by other American political leaders, such as Joe Barton and James Inhofe.[2] Metaphysical optimism may lead us either to deny the seriousness of the problem or to assert that there is another set of transcendent goods that are more important than the goods of this world. The latter concern can even lead to the perverse conclusion that climate change and environmental crises are portents of the end of times, which we should greet with apocalyptic hope. It is significant to note that nearly half of Americans believe that climate change is a sign of the Biblical apocalypse, according to a 2014 report by the Public Religion Research Institute and the American Academy of Religion (2014, 23–24).

Metaphysical optimism is an unreliable guide for ethical action. Should we pray and hope for the end of times—or should we hope against hope that God might prevent us from destroying the world? Those sorts of questions are irrelevant from the standpoint of humanistic, pragmatic meliorism. We certainly do need courage and a limited sort of hope. But we must be open to the possibility that catastrophes are not always averted and this is not the best of all possible worlds. We may have to revise our goals in light of changing circumstances. Nonetheless, we should hope that our choices and behaviours matter ethically. Whether we can prevent catastrophe is not clear. But we can certainly refuse to make things worse, while exhibiting virtues and doing our best to fulfil our duties within the exigencies of the moment.

## CONCLUSION

One source of obligation within virtue ethics can be found in taking an objective view of one's life as a whole—in asking yourself how you would like the narrative of your life to unfold (see Fiala 2009). Do we want to be remembered as egoists and hedonists who opted for short-term self-interest at the expense of others and future generations? Do you want to be remembered as Nero? Or as one of the musicians on the deck of the *Titanic*?

Of course, a serious pessimist may argue that this does not matter—since, if the threat of collapse is real, there may be no future generations for us to worry about and our narrative will go down with the ship. The problem of Nero's fiddle becomes acute if we believe that civilization will really collapse or that our very species will go extinct. Why sacrifice now, if there will be no recognizably human future? The problem of our obligations to nonhuman beings may be considered in much the same way. In the face of an imminent collapse of the Earth's ecosystem—in the face of the mass extinction and loss of biodiversity that some are predicting—it may be that we are all doomed together. In such a circumstance, one might conclude that it is not worth the sacrifice to care for endangered species. Who cares about rhinos and whales, when they are doomed anyway (along with the rest of the global mega-fauna, including ourselves)? Why care about future generations, if there is no hope that they will inherit a habitable Earth?

And here we discover the nihilistic nadir of hopelessness. It is possible to imagine a dark and despairing conclusion that holds that since we are all doomed, there is absolutely no point in taking proactive and protective steps. Such a null-point of despair can easily lead people to give up on practical steps and may even exacerbate environmentally damaging behaviours. If the rhino is going extinct anyway, then why not hunt the last of them for fun? If future generations are doomed, then why not use up the rest of the ground water for the needs of present generations? The risk of hopeless pessimism is a swift and nihilistic demise.

Morality requires, however, that we fend off despair. We can avoid the nihilistic conclusion by readjusting our hopes in light of changing circumstances. Short-term successes can be achieved. The rhino may go extinct, but perhaps we can save the elephant. We should avoid false dichotomizing that says 'either we repair the earth and return it to a pristine state—or we are all doomed'. Obviously, there are multiple possible outcomes within the broad range outlined by that dichotomy. And within that broad space of possibility, there is hope that some things we do can make a difference.

Thus, we should be careful about painting things with too broad of a brush. The challenge of future generations is a matter of degree and breadth of concern. Should we concern ourselves with the unborn seventh generation? Or should we concern ourselves with our own living children? Clearly, with regard to our own children, it would be immoral to give up hope and give in to eco-nihilism. Just as clearly, we don't really know what the seventh generation from now will want (or whether they will actually exist). Here again, a melioristic and pragmatic sensibility should prevail. We can take finite steps here and now to save rhinos and elephants and to preserve some goods for future generations. We cannot allow nihilism to cause us to give up hope in the short term.

Moreover, the concerns of environmental justice in the here and now are morally significant and within our power to resolve. Poor people will suffer more in the hot, crowded world of the Anthropocene. But we can act now to help them. To fiddle like Nero in the face of environmental injustice is obviously wrong when some—even a few—human beings need our immediate assistance. And finally, at the end of the day, we each have to live with ourselves. Stoic resolve and courageous virtue are ends in themselves. At some point, death comes to us all—to individuals, to civilizations and to species. We can choose how we respond to the face of calamity—just as we can choose how we respond to the facticity of death. Each action we take contributes to the story of our lives. While we may have to give up hope for a rescue and return to the pristine world of our ancestors, we cannot give up hopefulness as a positive disposition and virtue. Our actions matter—here and now—to ourselves, to suffering others and to those we love. We must do our best to do our work—to take appropriate action, give hope to others and behave properly—even if we are playing a requiem on the *Titanic*.

## NOTES

1. NPR. 'Is The Good Book A Good Guide In The Climate Change Debate?' November 18, 2010, http://www.npr.org/blogs/13.7/2010/11/18/131412508/is-the-good-book-a-good-guide-in-the-climate-change-debate; see also People for the American Way, 2011, 'The "Green Dragon" Slayers: How the Religious Right and the Corporate Right are Joining Forces to Fight Environmental Protection', http://www.pfaw.org/rww-in-focus/the-green-dragon-slayers-how-the-religious-right-and-the-corporate-right-are-joining-fo.

2. 'Texas Congressman Cites Noah's Ark As Evidence Against Climate Change'. *Mother Jones*, April 12, 2013, http://www.motherjones.com/environment/2013/04/texas-congressman-cites-noahs-ark-evidence-against-climate-change; 'Why Do Evangelicals Like James Inhofe Believe That Only God Can Cause Climate Change?' *New Republic*, January 29, 2015, http://www.newrepublic.com/article/120889/evangelical-james-inhofe-says-only-god-can-cause-climate-change.

## *Chapter Four*

# Hope, Self-Transcendence and Environmental Ethics

John Nolt[1]

Environmental ethicists often hold that organisms, species, ecosystems and the like have goods of their own. But, assuming that such goods exist, whether we ought to value them is controversial. Hence an environmental philosophy needs, in addition to an account of what sorts of values there are, an explanation of what, how and why *we* morally ought to value—that is, an account of moral valuing. This chapter presents one principle of such an account. Specifically, I aim to show that unless there are eternal goods (and maybe even if there are), we have a duty of self-transcendence toward nature—that is, a duty to value at least some of nature's goods as ends. It is, however, a duty owed not to nature, but to ourselves. It is justified by what I call an imperative of hope. The argument, in a nutshell, is this: we have a duty to ourselves to optimize (in a certain sense to be explained below) individual and collective human hope. This optimization requires self-transcendence toward entities whose goods are more diverse and enduring than any human goods. But unless there are eternal goods, such goods occur only in nature.[2]

That, of course, is just a sketch. Before filling in the details, I need to explain the concept of self-transcendence, distinguish it from identification, show that self-transcendence toward nature is possible and tease out various complexities of the concept of hope.

## DEFINITION OF SELF-TRANSCENDENCE

For the purposes of this chapter, self-transcendence is to be understood as follows: a person (*subject*) is *self-transcendent* toward an object if and only if that object is distinct from her and she values its good as an end.[3] The object might itself be a person, but it could be something else: a nonhuman animal, corporation, nation, etc. To value a good *as an end* is to value it noninstrumentally—that is, not merely as a means to other valued goods. To *value* it (in the sense relevant here) is to regard it as good and desire it.[4] Thus, for example (to put these pieces back together), someone is self-transcendent toward the birds in her neighbourhood if she regards their good as good in itself and desires it, not merely as a means to her own or others' goods, but as an end.

Self-transcendence is an intentional state. It can be directed toward a nonexistent object (as when someone loves a fictitious being) or toward an existing object that has no good or whose good it misconceives.[5] But in such cases it fails, because its aim is to value the authentic, not merely apparent, good of an objectively existing, not merely apparent, other. Where the object of self-transcendence exists and the good valued is the object's own (its *authentic* good), conceived with reasonable accuracy, then I call the self-transcendence *true*. And since this chapter's concern is exclusively with *true* self-transcendence, I will henceforth omit the qualifier 'true', except for occasional reminders.

Self-transcendence is fundamental to many forms of ethics. That there are duties of self-transcendence toward other people is widely acknowledged—though not, in general, by that description. The Golden Rule prescribes such duties, on the plausible assumption that we would have others value our good as an end. Kant's second formulation of the categorical imperative requires us to treat *people* (not their goods) as ends, but for Kant this means that we should value their goods (as defined by their rational ends) noninstrumentally.[6] Humanistic utilitarianism (whose aim is the aggregate good of humanity) is generally understood to require only that we promote that good, not that we value it, but surely valuing it (that is, self-transcendence toward humanity in aggregate) is instrumental to its promotion. For care ethics, obligations arise from relationships, but relationships of love, friendship and moral consideration all entail obligations of self-transcendence.[7]

Duties of self-transcendence toward sentient nonhumans are also widely recognized. Tom Regan's Respect Principle posits a duty self-transcendence toward all subjects of a life.[8] And Peter Singer, assuming duties of self-transcendence toward humans (whose good he understands as satisfaction of their interests), deduces via his Principle of Equality similar duties of self-transcendence toward all sentient beings.[9]

All such principles demand only *true* self-transcendence. We would not succeed in fulfilling their prescriptions by valuing the goods of nonexistent objects, or the goods of existent objects that do not have goods, or goods otherwise grossly misconceived.

## THE POSSIBILITY OF SELF-TRANSCENDENCE TOWARD NATURE

Environmental ethics asks whether there are duties of self-transcendence toward nature—though, again, not usually in those terms. But that question raises another: is self-transcendence toward nature even possible?

By 'nature', I mean those parts and aspects of the world that are neither human nor products of humanity. Nature includes, for example, all nonhuman, nondomesticated and nongenetically engineered living organisms and various functional aggregates of them—species, populations, ecosystems, etc. But it also includes geological, hydrological and meteorological, planetary and galactic systems and their components. While one can value any of these things—perhaps even value them somehow as ends—self-transcendence is *true* only if its object has a good of its own that we both value as an end and conceive reasonably accurately. That any nonliving thing—star, cloud, crystal, atom—has a good of its own is doubtful. True self-transcendence toward an object requires, moreover, an accurate conception of its authentic good, and it seems unlikely that we have any accurate conception of the authentic good of any nonliving thing. Self-transcendence toward nature, then, means primarily, if not exclusively, self-transcendence toward nature's *living* things.

Self-transcendence toward sentient animals is certainly possible. Indeed it is obligatory, given even the least controversial of what Regan calls 'direct duty views' (e.g., the view that we ought, for animals' sakes, to avoid cruelty to them) (Regan 2004, chapter 6). But although I too hold that we have obligations—including an obligation of self-transcendence—directly to sentient animals (my view is close to Singer's), I will not assume any such obligations for the purposes of this chapter. It is sufficient here to recognize that self-transcendence toward sentient animals is possible.

What about self-transcendence toward living but nonsentient entities: individual organisms, populations, species, ecosystems or the like? Nonsentient *organisms* do have goods of their own—or, at least, so I will assume. The case for this claim has been made by many environmental ethicists and need not be reiterated here.[10] We can, I will also assume, conceive the goods of nonsentient organisms reasonably accurately and value them as ends. Thus, for example, I may understand that the authentic good of a tree includes its growth and health, and I may value this good as an end. Of course, if I am to conceive that good accurately, I must not anthropomorphize. To value a tree's feelings, for example, is a factual mistake. But it is not a factual mistake to value—as defenders of trees sometimes do—a tree's well-being as an end. Given that nonsentient organisms have goods of their own, which we can conceive reasonably accurately and value as ends, self-transcendence toward them is possible. Whether biotic entities other than organisms have goods of their own is more controversial. I will not assume here that they do.[11]

Organisms generally, and especially nonsentient organisms, often have for human self-transcendence a certain interchangeability. We can value the good of an individual coyote or an individual tree. But we can also value the goods of organisms (e.g., the red-tailed hawks that soar over the hill of my home) without attending to their individual identities. In such cases, it is still the goods of *organisms*, not of anything so abstract as a species or population or ecosystem, that we value; yet those organisms are not always the same individuals from season to season, decade to decade, generation to generation. Something analogous, I think, occurs in what is called 'love for humanity': what one loves is people, not anything so abstract as the species *Homo sapiens*, but one loves them impartially, each like the others.

When I speak of self-transcendence toward *nature*, it is primarily this identity-indifferent valuing of the goods of organisms that I have in mind. Self-transcendence toward an individual coyote counts too, as does self-transcendence toward a species or ecosystem (if such a thing is possible), but these are, perhaps, less typical cases. Thus understood, self-transcendence toward nature is plainly possible.

The question with which this section began, however, was not whether it is possible, but whether it is a duty. A strong case can be made that we have duties of self-transcendence to sentient animals. But it is not clear that we have *any* duties to nonsentient entities. Elsewhere I have surveyed a range of arguments for duties to nonsentient entities and found none that succeeded (Nolt 2006). But I also noted this unrefuted possibility: that we have duties of self-transcendence toward, or identification with, natural entities that are owed not to those entities but to *ourselves*.

## SELF-TRANSCENDENCE DISTINGUISHED FROM IDENTIFICATION

With regard to identification, I now regard that suggestion as misguided. *Identification* with an object means taking its good as one's own.[12] This might be a mere metaphor for self-transcendence; that is, 'taking the object's good as one's own' might simply mean valuing the object's good as an end. But if understood more literally, it implies that harms or benefits to the object are ipso facto harms or benefits to me. Self-transcendence, by contrast, does not entail regarding benefit or harm to its object as benefit or harm to me. I may desire the goods of red wolves as ends, yet consider myself sufficiently distinct from them that I need not be benefited or threatened when they are. While there is consensus that we have duties of self-transcendence toward other humans, I see no reason to think that we have duties literally to *identify* with other humans, let alone with nonsentient beings.

I do contend, however, that we have a duty of *self-transcendence* toward nature, even toward nonsentient nature. This duty is grounded in a duty to ourselves to hope (or, equivalently, to avoid despair). To show this requires an articulation of the surprisingly complex concept of hope. That is the burden of the following two sections.

## THE DIMENSIONS OF HOPE

Hope is an intentional attitude of a person toward a state of affairs, which we may call its *object-state*. When George W. Bush hoped that the troops would find weapons of mass destruction, for example, the object-state of his hope was the troops finding the weapons. To *hope* for a state of affairs is to value it (regard it as good and desire it) and think it possible. (If we do not regard the object-state as possible, then our attitude is a mere wish, not a hope; if we do not regard it is good, then our attitude is one of desire, but not hope.) Hope is dispositional; one need not be conscious of it at a particular time in order to have it then.

Hope is, further, a matter of degree, assessable along various dimensions. There is first of all the strength of the person's desire for the object-state, which we may call the *strength* of the hope. Hope's desire tends to be stronger for the goods of those with whom we have more intimate relationships, and strongest for our own good. Where hope's desire is very small, the hope itself becomes ephemeral and insignificant; where it vanishes altogether, the hope vanishes as well, leaving (if anything) only expectation.

The strength of hope's desire should not be confused with three additional dimensions of hope that describe the person's often tacit and usually quite vague assessments of the object-state. These three dimensions are the anticipated goodness, likelihood and duration of the object-state. We cannot hope for something without thinking of it as in some sense good and having some notion, however indistinct or erroneous, of how good it is.[13] I may hope for an end to my cold (a good I take to be relatively small) or a cure for my cancer (a good I regard as much greater). Similarly, we also have some sense of how likely the hope's object-state is. At the very least, as noted above, we regard the object-state as possible. But often our sense of its likelihood is more specific. In hoping to find my lost watch, for example, I may regard success as very likely or not likely at all. Finally, hope involves some sense of the object-state's duration. I may hope for a good day tomorrow or for a good life in the coming decades.

The person's sense of the goodness, likelihood and duration of the object-state together constitute what I call the *magnitude* of the hope. We may think of magnitude as something like the product of the three—not the numerical product, of course (since the three factors are not definite quantities) but a rough analogue of it.[14] Thus a hope is great to the extent that the person regards its object-state as good, enduring and likely, smaller in proportion as her sense of the goodness, duration or likelihood decrease. The hope vanishes if her assessment of any of these three drops to zero.

Finally, some hopes are satisfiable and some are not. A person's hope is *satisfiable* (in the sense intended here) if its object-state is possible and her assessment of its goodness and duration are not greatly exaggerated or otherwise misconceived. If a hope overestimates or misconceives the goodness of its object-state, for example, then hoped-for good cannot occur. In that case, even if the object-state is realizable, the hope is still in an important sense unsatisfiable. If, for instance, I hope that the human population will plummet, my hope may not be satisfiable even if some great die-off occurs, if that is not the good I hoped it would be. Thus satisfiable hopes are hopes for goods that can be realized in more or less the way the hope conceives them. A satisfiable hope need not, however, be satisfiable in a person's lifetime. Moses' hope that Israel would enter the Promised Land, though satisfiable only posthumously, was nevertheless satisfiable.

I have so far mentioned only particular hopes (e.g., Bush's hope or Moses'). But we may also consider a person's *aggregate hope*—the totality of that person's particular hopes at a given time. A person's aggregate hope, too, has a magnitude, which we may think of (again with appropriate caveats for imprecision) as the sum of the magnitudes of each of her current particular hopes.[15] Thus we may say that a person has great hope if she hopes for many object-states that she assesses as very good, durable and likely.

So much for the dimensions of hope. In the next section, I will argue that we have a duty to uphold the magnitude of aggregate satisfiable hope, and in the four sections following it I will argue that that duty can be fulfilled (unless there can be satisfiable hopes for eternal goods) only by cultivating self-transcendence toward nature.

## THE IMPERATIVE OF HOPE

Hopelessness or dearth of hope is a form of suffering—namely *despair*. Thus, since (I assume) we have a moral duty to prevent and relieve suffering, we have a moral duty to uphold hope—or, equivalently, to avoid despair. It may seem silly to claim that we have a duty to do what is so clearly in our own interests. But we do, it is only because despair unfits us for service to others. This imperative of hope is also thoroughly communal. We ought to encourage hope in one another, and we as a society ought to cultivate it generally. But encourage and cultivate in what dimension—in magnitude or strength?

It seems clear that, other things being equal, despair decreases as hope's magnitude increases. That is, if I am in despair, my suffering will generally be relieved to the extent that I can hope for goods that I regard as greater or more likely or more enduring. If, for example, I am hungry and without hope of eating this day, my despair will be relieved somewhat if I learn that there a few edible tubers growing nearby (a relatively small good). It will be relieved more if I learn that aid trucks loaded with food will arrive within the hour (a greater good). Likelihood is also a factor. My despair will be slightly relieved if I learn that there is a small chance of aid, much relieved if I know that aid is nearly certain to arrive. Despair also decreases, other things being equal, as the anticipated duration of hope's object-state increases. If my prognosis for one more month of life is lengthened to a year, my despair will, other things being equal, decrease.[16]

The relationship of suffering to hope's strength (i.e., the strength of hope's desire) is, however, not so straightforward. Strong hopes of great magnitude can be exhilarating. But it is *not* true that the stronger the hope the less the suffering. Strong hopes for improbable goods (e.g., the hopes of the mothers of the 'disappeared') are often anguished. And the dispassionate, whose hopes are relatively weak (though not so weak as to evaporate), do not generally suffer thereby. Hope must have some strength in order even to exist and to matter to us. But increasing its strength seems to have little benefit in reducing despair. It is primarily aggregate hope's magnitude, then, and not its strength, that we must increase to relieve despair.

People ordinarily have various hopes that are in constant flux. Old hopes fade and new ones arise. Ideally, there is a kind of equilibrium, so that the magnitude of aggregate hope stays high enough to prevent despair. We despair when the magnitude of our aggregate hope becomes too low. Given a moral duty to prevent and relieve suffering, then, it follows that we, both individually and collectively, have a duty to maintain the magnitude of aggregate hope and to increase it when it is low enough to permit despair.

Of course, none of this is quite as simple as I have so far made it. Our duty to prevent and relieve suffering is, for one thing, imperfect. It needs to be understood with certain qualifications—something like this: we (both individually and collectively) ought to prevent or relieve suffering *insofar as this is reasonably possible and consistent with other obligations*. So our obligation to sustain hope must be similarly qualified. The wording of the qualification is not important for our purposes, so I won't attempt to get it exactly right, but we need to keep it in mind.

## THE IMPERATIVE OF TRUTHFULNESS

There is also this complication: not all hopes are equally worthy of encouragement. Unsatisfiable hopes, in particular, are morally dubious. For since hope takes its object-state to be both possible and good in the way the hope conceives it, unsatisfiable hopes are in a clear sense *false*. Morality, however, demands truthfulness; ordinarily, we should not be in the business of promoting false hopes. There may be situations so grim that false hopes are the only hopes available, and perhaps then false hopes are better than none. But such situations are the exception, not the rule. If lack of hope is the problem and if satisfiable hopes of substantial magnitude are available, then the solution is to encourage these satisfiable hopes first. Whether we should ever in addition promote false hopes is a question that need not be settled here.

Thus our individual and collective duty comes to this: to keep the magnitude of aggregate satisfiable hope high enough to avoid despair (or to increase that magnitude if it is not high enough), insofar as is reasonably possible and consistent with other obligations. Let's call this *optimizing hope*. Our duty, then, is to optimize hope.

It might be objected that increasing the magnitude of aggregate hope is not the only way to relieve the suffering of hopelessness—that one might do it, for example, pharmacologically. If so, there might not be a need to optimize hope. There are, it is true, drugs that can anesthetize us against despair—at least for a time. If a person had some psychological disability that rendered him incapable of hope, then drugs of that sort might be his only resort. But their use would be merely palliative, not curative. They would provide, not hope, but only endurance of hopelessness. Such a person would, moreover, be an extreme case. Despair is often due, not to a generalized incapacity to hope, but to adverse circumstances (and, as I am about to argue, to short-sighted valuing). And for this more typical despair there are genuine cures: change or revaluation of circumstance. Where cure is possible, palliative care (e.g., drugs) might still be temporarily indicated. But the proper goal is cure. We still have a duty to optimize hope.[17]

The most obvious cure, of course, is change of circumstance. If my hopes are low because I am penniless, then acquiring some capital may be the best way to raise them. But there are limits to what we can do; in the long run, as I will argue in the next section, we cannot optimize hope without self-transcendence.

## HOW SELF-TRANSCENDENCE SUSTAINS SATISFIABLE HOPE

We can best appreciate the role self-transcendence in sustaining satisfiable hope by considering what we lose in its absence. Imagine a competent adult who is completely devoid of true self-transcendence. Such a person violates duties of self-transcendence to others and so is morally deficient. But she is also deficient in satisfiable hope. For she will die, and her death will destroy the only thing (if any) whose authentic good she values as an end: herself.[18] Thus the magnitude of her satisfiable hope must tend to zero as her death approaches.[19]

If she were accomplished, she could, of course, even at moment of her death, hope for the survival of her works, fame, power or the like—and these might indeed survive her. Hence her *hope* need not dissipate with the approach of her death. But that hope would not be *satisfiable*. For nothing that occurs after her death can be a good *for her*,[20] and *ex hypothesi* she values as an end no authentic good of anyone or anything but herself. She cannot satisfiably hope for posthumous fame or the like as means to the authentic goods of others, for she does not value the authentic goods of others, nor can she satisfiably hope for them as means to her authentic goods, which cease with her death. She might hope for them as ends in themselves, but in that case her hope would still be unsatisfiable, for there is no intelligible way for them to *be* good, independently of benefits to anyone or anything. In sum, she cannot *satisfiably* hope for any posthumous goods because, given her lack of self-transcendence, they cannot actually be good in whatever way she might hope them to be. Thus as her death approaches, the magnitude of her satisfiable hope does indeed tend to zero. She must either take refuge in false hopes—a risky business, since there is always the danger that her true situation will become undeniable—or endure the increasing threat, if not the dominance, of despair.

But all of this follows only given an utter lack of true self-transcendence. True self-transcendence can increase the magnitude of satisfiable hope and extend the duration of its object-states beyond one's own death. To illustrate, suppose that the person of our previous example acquires a lover whose good she understands reasonably well and values as an end. She thus values as ends both her lover's good and her own. Provided that her lover could outlive her and that she recognizes this, she has thereby acquired the satisfiable hope, which she lacked before, that her lover may fare well in her absence.[21] Now, not all the authentic goods she values need come to an end with her death. The magnitude of her satisfiable hope need not, therefore, tend to zero as her death approaches. It will, however, almost certainly diminish. And if her lover dies first, her self-transcendence will cease and she will be in the same predicament as before.

She can, however, make her hope more sustainable by *broadening* her self-transcendence—that is, by directing it toward objects greater in quantity, diversity and/or duration. Imagine, for example, that she comes to value as ends the goods not only of her lover but of friends and family, children and grandchildren. Her death is then a relatively smaller loss of the authentic goodness she values, as long as her friends and family continue to fare well. The goods for which she hopes will probably endure longer and certainly be greater than if she valued only herself or only herself and her lover. The magnitude of her satisfiable hope is therefore greater still. Perhaps she can even approach death in the satisfiable hope that most of what she values will *not* end with her death. And if she broadens her self-transcendence still further (valuing as ends, say, goods of her community, humanity or nature), the goodness and duration of the goods she can satisfiably hope for may be still larger, as will the probability that at least some of her hopes will be satisfied. The magnitude of her satisfiable hopes may thus remain great even to the moment of her death.

In general, then, we can significantly increase the magnitude of our aggregate satisfiable hope in all of its dimensions simply by broadening self-transcendence.[22] Indeed, we can increase satisfiable hope even without improving our external situation—even when action is futile or impossible. And, if we *can* act, broadening self-transcendence, by giving us new goods to value, opens new possibilities of beneficial action.

Broadening self-transcendence does, it must be conceded, also make us vulnerable to greater loss. Among many hopes, many will fail. But as long as others replace them, we need not fall into despair. There is an ecological analogy here: by diversifying aggregate hope, we increase its sustainability. Hopes come and go, but aggregate hope remains.

Let's now take stock. I have argued that we have a duty to optimize hope and that the magnitude of satisfiable hope is sustainable through a lifetime only by self-transcendence—the broader the self-transcendence, the more sustainable the hope. Our duty, then, is to achieve and maintain broad self-transcendence. That, however, is nothing new. For, as I noted earlier, the standard forms of morality all prescribe fairly broad self-transcendence. They generally prescribe it, however, as a duty *to others*. What (if anything) is novel so far is the prescription of self-transcendence as a duty *to ourselves*, both individually and collectively—a duty grounded in the duty to optimize hope.

In the section titled 'The Imperative of Self-Transcendence toward Nature', I will venture a more novel conclusion, one directly relevant to environmental ethics: that unless we can have satisfiable hopes for eternal goods, we have a duty to ourselves (both individually and collectively) of self-transcendence *toward nature*. This conclusion is based on the claim, defended in the next section, that we cannot eliminate despair if the only goods we value as ends are those of human beings (individually or collectively) and of their artefacts.

## THE INADEQUACY OF HOPE FOR HUMAN AND ARTEFACTUAL GOODS

We have a duty to optimize hope—that is, to prevent or relieve despair (insofar as is reasonably possible and consistent with other obligations) by sustaining or increasing the magnitude of satisfiable hopes. But what sorts of hopes are satisfiable? To hope for an object-state is, as I have noted, to conceive it as good in a certain way. The hope is satisfiable only if that object-state is in fact good in that way. So the range of satisfiable hopes is constrained by what sorts of goods there are.

We can divide the candidates into four categories. There are, first of all, finite human goods, individual or collective. Individual goods include (but are by no means limited to) health, relief of suffering, friendship, love, freedom from oppression, etc. There may be *collective* human goods as well—such things as the flourishing of a nation or a culture. Secondly, there are the authentic goods of human artefacts. I here count, for example, domesticated animals as human artefacts. Clearly they have goods of their own. Genetically engineered organisms also fall into this category, as would robots, were they ever to become so sophisticated as to have goods of their own. Most artefacts, however (e.g., forks), do not have goods of their own and are good only as means to human goods. This section focuses exclusively on the finite goods of humans and artefacts. The third and fourth categories, natural and eternal goods, are the subjects of the sections titled 'The Imperative of Self-Transcendence toward Nature' and 'Eternal Goods', respectively. My main claim in this section is that although we can have satisfiable hopes, including hopes satisfiable only posthumously, for the goods of humans and their artefacts, these hopes by themselves cannot eliminate despair.

Thoreau penned his famous line 'The mass of men lead lives of quiet desperation' (Thoreau, in Krutch 1981, 111) in an age of optimism. He was not thinking primarily of the manifest hopelessness of old age, poverty or oppression, but of the mostly suppressed despair that haunts even the young, rich and comfortable. His observation has not lost validity with time. Witness the proportion of contemporary Americans on antidepressants.

Hope for merely human goods (goods of self, family, friends, organizations) sustains many (though by no means all) people through much of their lives, but toward the end it often falters. Concerning the value of self-transcendence toward family in the face of his own death, Leo Tolstoy wrote:

> But my family—wife and children—are also human. They are placed just as I am: they must either live in a lie or see the terrible truth. Why should they live? Why should I love them, guard them, bring them up, or watch them? That they may come to the despair that I feel, or else be stupid? Loving them, I cannot hide the truth from them: each step in knowledge leads them to the truth. And the truth is death. (Tolstoy, in Bayley 1978, 681)

Unsatisfied with self-transcendence toward family, Tolstoy eventually turned, as many do, to hopes for eternal goods. If self-transcendence toward other people, whose reality is manifest, sufficed to prevent despair, it would be less easy to account for the popularity of faith in things unseen.

This is not to deny that there is satisfiable hope to be gained, even at life's end, from self-transcendence toward family or other people generally. The point is merely that such hope is for many people insufficient to eliminate despair. Maybe by training or self-discipline we might come to invest nearly all our hope in others and so lose little at the prospect of our own death. Yet even that would not always be sufficient. The others for whose good I hope might die before me—and in any case, as Tolstoy frets, they too will die within decades. Moreover, since human relationships are fragile, we might become estranged, and so I might lose hope for them before they died. Much the same can be said for any human organization or project that I might value. I cannot know whether I will be one of those for whom human hopes fail. And even if I am not, it is clear that for people collectively despair cannot be avoided by merely human hopes.

Maybe what we need, then, is a generalized hope for all humanity. But that is probably beyond the capability of many, maybe all, people. We readily develop hopes strong enough to matter to us only for what we know fairly intimately. It is difficult even to comprehend, much less to hope for, the goods of billions of people. Humanity is, moreover, in many respects maddeningly foolish, short-sighted and self-destructive. We may easily despair even of the prospects of all humankind.

As for artefacts with goods of their own, these are generally not significant objects of hope. Perhaps in the far future we will invest great hope in our silicon offspring[23] or in the brave new products of our genetics laboratories, but today the role of the goods of artefacts in human hope is so small as to be for our purposes negligible. I will henceforth neglect it.

## THE IMPERATIVE OF SELF-TRANSCENDENCE TOWARD NATURE

Self-transcendence toward humans and their artefacts cannot, then, eliminate all despair. But, it will be objected, neither can self-transcendence toward nature. Quite so. But in this section, I will argue that self-transcendence toward nature combined with self-transcendence toward humans makes possible satisfiable hope of significantly greater magnitude than does self-transcendence toward humans alone. Since our duty is to optimize hope, and since hope for the goods of humans cannot prevent all despair, it follows that we have a duty of self-transcendence toward nature.

Of course, if we can have satisfiable hopes for eternal goods, hopes for anything finite may be superfluous, since the former alone may suffice to eliminate despair. But for the remainder of this section, I will assume for the sake of argument that there are no eternal goods. I'll discard that assumption in the section titled 'Eternal Goods'.

My immediate aim, then, is to show that self-transcendence toward nature combined with self-transcendence toward humans makes possible greater satisfiable hope than does self-transcendence toward humans alone. The idea is straightforward: since nature's goods combined with human goods are more diverse and almost certainly more enduring than human goods, if our aggregate satisfiable hope is directed toward nature as well as humans, then its object-states can be richer in value, more probable and more enduring than if it is directed toward humans alone; hence, its magnitude can be greater.

Consider, first, the value of the object-states. It is generally held that the authentic good of any nonhuman organism is of less value than the authentic good of any human. Let's grant that assumption. Still, the good of a human plus the good of a coyote (or even a tree) is greater than the good of the human alone. Hence if we are self-transcendent toward not only humans, but nature as well, we can satisfiably hope for greater goodness than if we are self-transcendent only toward humans.

One might, I suppose, object that each person is capable of only so much self-transcendence and no more, so that it is best to invest our limited share entirely in those objects whose goods are greatest: humans. But this objection assumes that (1) there are no natural goods greater than human goods, (2) our ability to value is completely inelastic and (3) the only relevant factor in deciding what we ought to value is degree of goodness. (1) is not obviously true—even assuming, as we are, that the authentic good of any nonhuman organism is of less value than the authentic good of any human—for natural goods might in aggregate be greater than human goods. I know of no good evidence for (2). And (3) is plainly false, for not only degree of goodness but also its likelihood and duration are relevant in choosing what to value. To these two additional factors we now turn.

Just as the goodness of hope's object-states is greater with self-transcendence toward both humans and nature than it is with self-transcendence only toward humans, so too is the likelihood of these object-states. This is largely a matter of diversity. The broader my self-transcendence, the more likely it is that at least some of the goods I value as ends will be realized. But it is also due in part to the interchangeability of natural organisms that I noted earlier. I do not value the hawks that soar above my hill in the same way that I value the people who matter to me (i.e., primarily as individuals). It suffices for my hope that their like will fly here generation after generation. The survival to any particular date of these hawks, considered without regard to their individuality, is, no doubt, more probable than the survival of any or all of the individual humans whom I love.

It is also likely to be of greater duration. Even individual organisms (e.g., the oaks and hickories on my land) may yet survive for several centuries. But red-tails have probably been soaring over this hill for tens, if not hundreds, of thousands of years. My hope that they will continue to flourish long into the future may well be satisfiable. Of course, I have long-term hopes for humans and certain of their institutions as well. But the two kinds of hopes are not mutually exclusive. And since it is quite likely that nature will outlast humanity, self-transcendence toward nature may thus greatly increase the duration of the object-states of satisfiable hopes.

But what if (as seems likely) the most enduring of nature's creatures are something nasty—like cockroaches? Would it be best, then, to become self-transcendent toward cockroaches?[24] Probably not. The first mistake here is to consider only one or two of the dimensions of hope's magnitude—the duration, and perhaps also the probability, of the object-state—and ignore goodness. That cockroaches have a good of their own, I do not doubt. But surely we may satisfiably hope for greater, if less enduring, natural goods. Moreover, because (let us grant) cockroaches are creepy and dirty, their good is to a certain extent incompatible with ours—at least when they inhabit our floors and walls. Self-transcendence toward cockroaches is therefore difficult or impossible for most people—and hence not to be widely recommended. Fortunately, nature offers to our hope many more congenial alternatives. We need not be self-transcendent toward all of it, but given the current threats nature, we will best preserve satisfiable hope by being self-transcendent toward a good bit of it.

To summarize: by self-transcendence toward nature as well as humans, we can reasonably achieve aggregate hope that is significantly greater in all three dimensions of magnitude—the goodness, likelihood and endurance of its object-states—than by self-transcendence toward humans alone. Therefore, since self-transcendence toward humans cannot eliminate despair, we (both individually and collectively) can optimize hope (i.e., optimally resist despair) only by self-transcendence toward both humans *and* some significant portion of nature. We have a duty to optimize hope. Thus, in particular, we have a duty of self-transcendence toward at least some aspects of nature.

But is this really a *duty*? The point, after all, is to optimize hope, not to maximize it. And we optimize hope if we merely prevent despair. Some people manage to sustain hope throughout their lives by self-transcendence, not toward nature, but merely toward other people and/or institutions. Isn't that, then, enough? Why posit a general duty of self-transcendence toward nature, when some people, at least, can optimize hope without it?

The premise of this objection is plainly right; some people do escape despair by satisfiable hope for merely human goods. None of us, however, can be sure of doing so. Hence the duty stands, for it derives not from the *actual* consequences of our valuings, but from their *likely* consequences. An analogy may be helpful here. Some people who eat badly never become ill in consequence. Yet still we owe it to ourselves and our families to eat healthfully, because failure to do so makes us substantially more likely to suffer and to become a burden to others. Likewise, some who are not self-transcendent toward nature never despair. Nevertheless, we owe it to ourselves and our communities to seek such self-transcendence, because failure to do so makes us substantially more likely to despair and to become a burden to others.

Still, it might be objected, we need not worry about despair now, while things are going well. When tragedy or old age arrives, then we can learn solace in nature. This strategy, however, will fail; for we cannot become self-transcendent overnight. To serve as a firm foundation for hope, self-transcendence requires long cultivation. Our duty, then, is not to become self-transcendent toward nature occasionally, as needed to prevent despair, but individually and collectively, to cultivate self-transcendence toward nature throughout our lives.

One further objection may arise at this point. I have taken it for granted that the objects of self-transcendence toward nature are living things. But given that living nature is itself so fragile, why not pin our hopes on something more permanent—something that will survive even the demise of all life? Why not hope, say, for the endurance of the stars, perhaps even for the continued reign of the laws of physics?

I have neglected such hopes primarily because I am not convinced they are satisfiable.[25] I do not, of course, doubt that the hoped-for object states can occur. The stars and the laws of physics will, I assume, last a very long time—probably longer than life itself. But, as I explained above, a person's hope is satisfiable in the relevant sense only if her assessment of the object-state's goodness is not greatly exaggerated or misconceived. I do not see how in the absence of any benefit to any living thing the endurance of the stars or of the laws of physics could be good in *any* sense, much less in a sense for which we could satisfiably hope.

But there is no need to be dogmatic on this point. If there is an intelligible sense in which, for example, the mere endurance of the stars or the laws of physics is good for them, or for some other entities, then self-transcendence toward these entities is possible. But in that case, it is merely another variety of self-transcendence toward nature and hence poses no problem for the account given here.

Still, even the most durable objects of satisfiable hope must eventually come to an end, if only by the collapse or heat death of the universe. But such an ultimate demise need not nullify our hopes for the interim, and between us t may lie an expanse of time so vast as to be for the purposes of human hope barely distinguishable from eternity.

## ETERNAL GOODS

What, then, of eternity itself? We arrived at the conclusion of the section titled 'The Imperative of Self-Transcendence toward Nature' (that we have a duty to cultivate self-transcendence toward nature) under the hypothesis that there are no eternal goods. Dropping, now, that hypothesis, we are obliged to qualify our conclusion as follows: *unless there are eternal goods*, we have a duty to cultivate self-transcendence toward nature.

So what if there *are* eternal goods? In that case wouldn't self-transcendence toward nature—or, indeed, toward any finite object—be superfluous? There are three issues here: (1) whether eternal goods exist; (2) if so, whether there can be satisfiable hopes for them; and (3) whether affirmative answers to both (1) and (2) would make self-transcendence toward finite objects superfluous.

To address (1) is beyond my competence. Regarding it, I will note only that while the existence of human and natural goods is obvious (health, for example, is a good for all organisms, including humans), evidence for eternal goods is more tenuous. But that does not settle the matter. Leaving it, then, unsettled, let's suppose for the sake of argument that eternal goods do exist.

Given that assumption, the second question is whether we can have satisfiable hopes for these goods. Hope, of course, is easy, but *satisfiable* hope requires that its object-states not be grossly misconceived. From the sheer variety of extant hopes for eternal goods, it seems probable that those goods are often grossly misconceived. Suppose, however, that this is not always the case and that some hopes for eternal goods are satisfiable. Still, such hopes may not be available to everyone. If, for example, the only eternal good is impersonal union with the godhead, as some Hindus believe, then satisfiable hopes for eternal goods may not be possible for those Muslims or Christians whose hopes demand *personal* immortality. In that case, even though eternal goods exist, the only *satisfiable* hopes of the latter are for finite goods.

Suppose, however, that satisfiable hopes for eternal goods are available to everyone. There remains issue (3): whether such hopes would render self-transcendence toward finite objects superfluous. In one sense, certainly not—since, as we saw earlier, self-transcendence toward other humans, at least, is a moral duty. But the point of the question is whether, given that some hopes for some eternal goods are satisfiable, self-transcendence toward finite objects is superfluous *as a source of satisfiable hope*. That depends in part on whether those eternal goods are distinct from the finite ones. It is conceivable that nature's goods are also in some sense eternal—that, for example, nature is most adequately understood, as Spinoza puts it, *sub specie aeterni* (Spinoza 1992, 93 [Part II, Prop. 44, Cor.2]). If something like this is true, then self-transcendence toward nature, far from being superfluous, is ipso facto self-transcendence toward objects whose goods are eternal.[26]

Suppose, however, to add yet another hypothesis to our growing list, that eternal goods are distinct from finite, worldly goods. In fact, take the extreme case of personal immortality replete with rich and unending goods for oneself (heaven, perhaps, or the disembodied contemplation of Platonic forms). Wouldn't that make self-transcendence toward nature (indeed self-transcendence of any sort) superfluous as a source of sustainable, satisfiable hope? Still, the answer is not obvious. For one thing, it is not clear that we ought to relinquish hopes that we have good reason to believe are satisfiable in favour of hopes for which we have less evidence, even if the latter really are satisfiable. Hopes for eternal goods are sometimes overwhelmed by doubt, especially in times of great loss. Given the duty to optimize hope, it might be best to cultivate both.

## AN APPARENT INCONSISTENCY

Thus even if there are eternal goods, we might still have a duty to cultivate self-transcendence toward nature. Let's assume for the remainder of this paper that we do have such a duty. Still, my account of this duty may seem inconsistent. For it is a duty *not* to nature, but to *ourselves*. My argument is consequentialist and anthropocentric, grounding this duty in a more general duty to prevent and relieve human suffering. And yet my conclusion—that we ought to value at least some of nature's goods as ends—is anything but anthropocentric.[27] It might, then, seem that I have contradicted myself, for if the ultimate aim of valuing nature's goods is to avoid human suffering, then those goods are *not* valued as ends, but merely as means to human ends.

We must distinguish, however, the authentic goods of nature (which are possible states of natural creatures) from the valuing of them (self-transcendence, an intentional state of the human being). It is the latter—the state of the human being—that in my account functions as a means to the end of preventing or relieving human suffering. The former, the goods of nature themselves, are in self-transcendence valued as ends, not as means to human ends. We may become self-transcendent for own benefit, but the benefit depends on valuing as ends goods that are *not* ours. Our self-interest, if it is to achieve its aim, must transcend itself, producing genuine desire for the authentic good of others, so that ultimately we desire as ends both our good and theirs. There is no inconsistency in that.

Still it may seem cynical or disingenuous to choose what to value strategically for its ability to promote hope—especially if that is our sole reason for valuing, and still more especially if the only hope we seek to promote is our own.

But nothing I have said implies that the enhancement of hope should be our sole reason for valuing. On the contrary: valuing in order to enhance hope should augment, rather than replace, our valuing for more familiar reasons. And there is nothing inherently selfish, cynical or disingenuous about cultivating, both individually and collectively, a virtue (broad self-transcendence) that renders us less susceptible to despair and hence better able to serve others.

## DEFENDING NATURE AS A SOURCE OF HOPE

I have given a partial account of moral valuing—of what, how and why we ought to value. A complete environmental ethic would do more; it would tell us how we ought to act. This final section adumbrates a few features of a more complete environmental ethic by considering briefly what else my account does or does not entail.

Let's begin with the 'does nots'. Nothing that I have said implies either that the duty of self-transcendence is the only duty we have with regard to nature (it obviously is not) or that the duty to optimize hope is the only justification for it. Though both duties, as I have argued, are owed to ourselves, not to nature, and though I have expressed doubt that there are duties owed directly to nature, nothing I have said implies that there are no such duties.

Indeed, it might seem that that the opposite is true—that the duty to value nature's goods as ends that I have advocated would entail, for example, a duty directly to nature to defend it when those goods are threatened. But I know of no plausible premises that when added to the inference would validate that inference:

> We have a duty to ourselves to value nature's goods as ends.
> Therefore, we have a duty to nature to defend it.

Our duty to optimize hope does, however, imply a duty *to ourselves* to defend nature. For in order to optimize hope, we must sustain the conditions for hope's optimization. One of these conditions is the preservation of nature; for to the extent that nature is degraded, the object-states of satisfiable hope diminish in goodness, likelihood or duration, and hope itself becomes less sustainable. We ought, if possible, to prevent this degradation. But to prevent it we must defend nature.

There are, of course, many other reasons for defending nature. All the usual anthropocentric arguments (e.g., nature as natural capital, nature as locus of beauty, nature as recreational resource, etc.) support the same conclusion. But even if all other justifications were to fail, we would still have an obligation to defend nature grounded in our duty to sustain possibilities of satisfiable hope. My account of moral valuing implies, moreover, something that purely humanistic theories do not: a duty to value nature's goods as ends, not merely as means.

That duty, together with the fact that we can most readily value that with which we are intimately acquainted, implies further that we ought to cultivate, insofar as is reasonably possible, intimate acquaintance with many aspects of nature. Environmental education that is not merely of the classroom is therefore an individual and collective responsibility. One moral aim of an environmental education should be to broaden self-transcendence so as to enlarge hope. While self-transcendence toward all nature, like self-transcendence toward all humanity, is an unattainable ideal, it is nevertheless an ideal, and environmental education should move us toward it.

The obligations to know, value and defend nature for which I have argued are obligations owed to ourselves—but not just to ourselves individually. They are also, as I have insisted, communal. I will conclude by indicating the scale of the relevant community, for that tells us something about these obligations' importance. Because the degradation of nature is often irreversible (extinction is an obvious instance) or reversible only in the very long term, the community of humans for whom we ought to defend nature (and hence preserve possibilities of satisfiable hope) includes not only us and our contemporaries, but all posterity.[28]

## NOTES

1. 'Hope, Self-Transcendence and Environmental Ethics' by John Nolt was originally published in *Inquiry: An Interdisciplinary Journal of Philosophy* 53 (2): 162–82 (2010). Reprinted by permission of Taylor & Francis Ltd., www.tandfonline.com.

2. The outlines of this argument were first suggested to me by a passage near the end of John O'Neill's rich and insightful paper, 'The Varieties of Intrinsic Value', *Monist* 75 (2) (1992), reprinted in A. Light and H. Rolston (2003), *Environmental Ethics: An Anthology* (Oxford: Blackwell), pp. 131–42. In this passage, O'Neill sketches this plan for justifying an environmental ethic: 'The most promising general strategy would be to appeal to the claim that a good human life requires a breadth of goods. . . . The ethical life is one that incorporates a far richer set of goods and relationships than egoism would allow . . . the recognition and promotion of natural goods as ends in themselves involves just such an enrichment" (139–40). In this chapter I combine O'Neill's strategy with the ideas that hope is a central requirement of a good human life and that it is sustainable only by self-transcendence.

3. This definition is in two respects weaker than that of Partridge's 'Why Care About the Future?' in E. Partridge, ed. (1981), *Responsibilities to Future Generations* (Buffalo: Prometheus Books). According to Partridge, a person is fully self-transcendent when '(a) he regards something other than himself as good in itself and (b) when he desires the good and endurance of this "something else" for its own sake, apart from its future contingent effects on him' (208). First, I omit part (a). To love (and hence be self-transcendent toward) someone, one does not need to think of that person as good in herself. We can love a person simply by wanting the best for her, without having any notion of whether she is good in herself. Second, I omit 'and endurance' because we may be self-transcendent toward an incurable sufferer for whom we

think it would be a mercy to die. With these two omissions, Partridge's definition is equivalent to mine.

4. This stipulative definition merely explains how I use the term 'value' in this chapter. One can, of course, esteem something as a good and yet not desire it, and we may think of that as 'valuing' in a sense. But it is not the full-blooded sort of valuing inherent in a self-transcendence that gives rise to hope, which is the sort of valuing that concerns me here.

5. When I say that an object is nonexistent, I intend to speak tenselessly. Thus I hold that self-transcendence may be true even if directed toward entities that do not exist at present, provided that they will exist. (Self-transcendence toward entities that no longer exist is, I suppose, also possible, but not of any interest here.)

6. 'For the ends of any person, who is an end in himself, must as far as possible also be my end, if that conception of an end in itself is to have its full effect on me' (Kant 1959, 430).

7. See, for example, chapters 5 and 6 in Tong 1993.

8. The principle is: '*treat those individuals who have inherent value in ways that respect their inherent value*' (emphasis in original) (Regan 2004, 248).

9. The principle is 'the interests of every being affected by an action are to be taken into account and given the same weight as like interests of any other being' (Singer 1990, 5). To give an interest weight is to value its fulfilment not merely as a means, but as an end.

10. Classic sources are Rolston 1988, chapter 3; and Taylor 1986, chapter 2. See also the exchange between Rolston and Ernest Partridge in Pojman 1997, 81–92; O'Neill in Pojman 1997, 137–38; and Nolt 2009.

11. Elsewhere, however, I have argued that they do. See Nolt 2009. For doubts about ecosystems, see Cahen 1988, reprinted in Light and Rolston 2003.

12. Arne Naess, for example, defines identification as a condition in which '*the interest or interests of another being are reacted to as our own interest or interests*' (emphasis in original) (in Tobias 1984, 261). Similarly, Warwick Fox in *Toward a Transpersonal Ecology* (1995, 249–50), writes that in personally identifying with various entities 'we experience these entities as part of "us", as part of our identity. An assault upon their integrity is an assault upon our integrity'.

13. We may of course hope for things that are not good—a cigarette, for example, or the suffering and humiliation of those whom we despise. But insofar as we hope for such things they or some aspect of them will seem good to us. Otherwise our attitude is mere desire.

14. This is an oversimplification, accurate only if the expected probability and goodness do not vary over the time envisioned by the hope—that is, only if they are constants. For a more explicit account of what I have in mind, I offer the following mathematical model. (It is only a model, and hence not to be taken too seriously, since it unrealistically assumes that the anticipated goodness, likelihood and duration of the object-state have numerical values.) Let $p$ be some moment (e.g., the present) at which one is hoping for an object-state $s$, and let $t$ be any later moment. Then the *momentary expected value* $v\ s(t)$ of $s$ at $t$ is the assessed goodness of $s$ at $t$ multiplied by the assessed likelihood of $s$ at $t$, both estimated from the perspective of $p$. Clearly $v\ s(t)$ can be nonzero only for times $t$ at which one hopes that $s$ will occur. Then the magnitude at $p$ of hope for $s$ is given by:

$$\int_p^{\infty} v_s(t)dt$$

15. This assumes that the hopes are individuated in such a way that they do not 'overlap'.

16. Assuming, of course, that I want to live.

17. It may be objected that there is a third possibility: a state of detachment, without either hope or despair, such as some Buddhists or Stoics advocate. This is not the place for a digression on detachment. My reply can only be brief: such detachment is typically not a cessation of hope, but a replacement of its usual object-states by object-states of a different kind. (A Stoic may hope to live a life of integrity. A Buddhist may hope for universal enlightenment.) Absolute lack of hope for anything is not, so far as I can see, a desirable form of detachment.

18. Believers in personal immortality should add to this example the stipulation that the only authentic goods the person values (if any) are the *mortal* goods of her *mortal* self. Of course if

there is no personal survival or if there is but she thoroughly misconceives its goods, then no such stipulation is necessary. Whatever she values regarding her immortal self is in that case not an authentic good at all. I will say a bit more about immortality in the section entitled 'Eternal Goods'. There is also a sense in which a relational self might survive death. But relational selves are self-transcendent and the person of this example is not.

19. I have formulated this in my own way, but I owe the idea to peripatetic conversations with John Hardwig that took place a couple of decades ago.

20. John O'Neill (1993) argues, to the contrary, that what happens after death can benefit or harm us. But his argument, while cogent, assumes our self-transcendence and hence is not applicable to the case at issue here. The actions of future generations may, he asserts, determine whether our projects succeed or fail; thus, they may benefit or harm us. But this is true because those projects aim for communal goods—such as, for example, the preservation of institutions or the progress of science. Someone who values as ends no authentic goods apart from her own has no projects aimed at communal goods. See also Levenbook 1984.

21. She has acquired the hope because (1) given her self-transcendence, she values the realization of her lover's good even after her death and (2) believes this to be possible, which two conditions are sufficient for her hoping that her lover fares well after her death. This hope is, moreover, satisfiable, because (we have assumed) it is possible that that her lover will outlive her and (since her self-transcendence is true) she conceives the lover's authentic goods fairly accurately.

22. Provided, of course, that the goods that we hope for are possible, but this condition is usually not difficult to meet.

23. I have in mind here something like von Neumann machines—robot probes designed to colonize the universe. See, for example, Tipler 1994, chapter II.

24. Thanks to my daughter, Jenna Nolt, for raising this objection.

25. There is also the question of whether people's desires for such things can be made strong enough to support much hope, but that is an empirical matter that need not be decided here.

26. Erazim Kohák, for example, advocates what I take to be such a view in *The Embers and the Stars* (1984).

27. An anthropocentric view is one that sees all value as value for human beings.

28. I wish to thank Elijah Weber and Baylor Johnson for valuable comments on a portion of this chapter, which I presented at the fifth annual joint meeting of the International Society for Environmental Ethics and the International Association for Environmental Philosophy at Estes Park, Colorado, 27–30 May 2008. This chapter has also benefited from the comments of two anonymous referees for *Inquiry: An Interdisciplinary Journal of Philosophy*.

## *Chapter Five*

# Optimizing Hope

## *A Response to Nolt*

Trevor Hedberg

Principles of environmental ethics are often supported by a pattern of reasoning that Holmes Rolston (1988) describes as a 'transition from *is* to *good* and thence to *ought*' (231). The first step is a connection between fact and value. One might, for instance, claim that entities with certain features or properties possess a certain value. The second step is to claim that the value present generates moral obligations. Here is a simplistic illustration:

1. All sentient creatures have interests.
2. We ought to respect all creatures that have interests.
3. Therefore, we ought to respect the interests of sentient creatures.

The first premise represents the move from *is* to *good*: it claims that a certain value (i.e., having interests) is shared by all entities with a certain attribute (i.e., sentience). The second premise represents the move from *good* to *ought*: it claims that we are morally required to treat all creatures with this value (i.e., having interests) in a particular way. Combining these premises, we arrive at a principle of environmental ethics—a moral imperative to respect the interests of sentient creatures.

John Nolt (2006, 2009) has written about both these argumentative maneuvers in his previous work, but I will focus only on the second one: the move from *good* to *ought*. In Nolt's (2006) survey of *good*-to-*ought* moves, he concludes that *good*-to-*ought* manoeuvres 'do not come cheap' and that justifying them 'remains one of the central open problems of environmental ethics' (374). But Nolt was not content to leave the issue unresolved. His 'Hope, Self-Transcendence and Environmental Ethics' is an attempt to defend a unique move from *good* to *ought*. Proceeding from the assumption that living organisms—both sentient and nonsentient—have goods of their own, Nolt argues that these goods generate moral obligations for human beings because we have a general duty to become self-transcendent toward living things in nature. Self-transcendence refers to an intentional state of valuing the good of some object other than yourself as an end (p. 44, this volume). Thus, if successful, the argument supports a partial biocentrism—the view that we should regard a significant portion of nonsentient life (as well as sentient life) as having direct moral standing.[1] Intriguingly, the starting point for Nolt's argument is actually anthropocentric: it begins from the general duty to prevent and relieve *human* suffering. This duty generates an obligation to maintain enough satisfiable human hope so as to avoid despair; Nolt refers to this obligation as the duty to 'optimize hope' (p. 50, this volume). He then argues that optimizing hope generally requires achieving self-transcendence toward nature.

Nolt's argument, which I will hereafter call the Hope Argument, is novel and thought-provoking, but nonetheless, I contend that it is unsuccessful. The purpose of this chapter is to highlight the shortcomings of the Hope Argument and propose an alternative view regarding how much self-transcendence toward nature we should seek. I begin my critique by outlining the central claims of the Hope Argument and identifying the premises that can be challenged. Afterward, I consider some objections that might be raised to these premises. Although the Hope Argument can withstand some of these criticisms, I argue that there are three objections that it cannot fully overcome. First, the general duty to optimize hope can only be understood as a collective duty. Given the variance in the psychologies of individual people, individuals cannot each have duties to seek self-transcendence toward living things in nature. Second, it is not clear that self-transcendence toward living things in nature increases our probability of optimizing hope: given that it makes us vulnerable to devastating losses, it may actually take us *further* from optimizing hope. Third, the amount of hope generated through self-transcendence toward nonsentient entities is minute compared to the amount of hope generated by self-transcendence toward sentient creatures and other people. Thus, the broad self-transcendence that Nolt advocates may not make a significant contribution toward optimizing hope. I conclude my critique by defending an alternative view—namely, that self-transcendence should only be directed toward sentient life. I argue that this narrower self-transcendence is enough to optimize hope and better accords with other moral values we hold.

## UNPACKING THE HOPE ARGUMENT

The Hope Argument is complicated, and while readers can consult the previous chapter for its details, it will be helpful to recast the argument in a more digestible form. We can then get a sense of where its weaknesses reside. First, however, we should review the crucial terms and philosophical concepts that Nolt introduces as he presents the Hope Argument:

- *True self-transcendence:* a kind of self-transcendence that occurs when the object of self-transcendence exists, the good of the object is conceived with reasonable accuracy and the good valued is the object's own (p. 44, this volume)
- *Nature:* those parts and aspects of the natural world that are neither human nor products of humanity (p. 45, this volume)
- *Hope:* an intentional attitude of a person toward some state of affairs such that the person regards it as good and desires it while also believing it is possible (p. 47, this volume)
- *Object-state:* the state of affairs for which one hopes (p. 47, this volume)
- *Strength of hope:* strength of a person's desire for the object-state to be realized (p. 47, this volume)
- *Magnitude of hope:* the (nonnumerical) product of a person's sense of an object-state's goodness, likelihood and duration (pp. 47–48, this volume)
- *Aggregate hope:* the totality of a person's particular hopes at a given time (p. 48, this volume)
- *Despair:* a form of suffering characterized by a dearth of hope (p. 48, this volume)
- *Satisfiable hopes:* hopes in which the desired object-state is realizable (p. 48, this volume)
- *Unsatisfiable hopes:* hopes in which the desired object-state is not realizable (p. 48, this volume)

Having reviewed the terms that play a role in the Hope Argument (and in my upcoming critical remarks), we can now reconstruct its many premises and sub-conclusions. Nolt abbreviates 'true self-transcendence' as simply 'self-transcendence', and I follow the same convention in my presentation and discussion of the argument.

1. We (both individually and collectively) ought to prevent and relieve human suffering, insofar as this is reasonably possible and consistent with other obligations.[2]
2. Despair is a form of suffering.
3. Therefore, we ought to avoid despair. (1, 2)

4. We can only avoid despair by maintaining a satisfactory level of aggregate hope.
5. Therefore, we ought to maintain a satisfactory level of aggregate hope. (3, 4)
6. Hopes are either satisfiable or unsatisfiable.
7. Cultivating unsatisfiable hopes is generally impermissible, particularly when satisfiable hopes are available.
8. There are many states of affairs for which one can satisfiably hope.
9. Thus, we ought to keep the magnitude of aggregate *satisfiable* hope high enough to avoid despair. (5–8)
10. The satisfactory magnitude of satisfiable hope is sustainable through one's lifetime only through self-transcendence.
11. Self-transcendence tends to be more sustainable the broader it is.
12. Despair is so common and difficult to avoid that we can reliably maintain hope through our lifetimes only through broad self-transcendence.
13. Hence, we ought to maintain broad self-transcendence. (9–12)
14. It is possible to achieve self-transcendence toward four types of goods: finite human goods, goods of human artefacts, natural goods and eternal goods.
15. Human goods and the goods of human artefacts are usually not enough when we near the end of our lives or otherwise encounter great loss or stress.
16. There are no eternal goods. (conditional proof assumption)
17. Therefore, maintaining broad self-transcendence requires achieving self-transcendence toward nature. (13–16)
18. Hence, if there are no eternal goods, then we ought to cultivate broad self-transcendence toward nature. (16, 17)

It should be obvious from this reconstruction that the Hope Argument contains a lot of premises. Since it only takes one false premise to make an argument unsound, one might think that an argument with this many steps is bound to go wrong somewhere, but pinpointing exactly where such a misstep occurs is a fairly tough task.

## CRITIQUING THE HOPE ARGUMENT

The Hope Argument begins from a general duty to reduce the overall human suffering in the world, which is widely regarded as one of the most basic moral imperatives. Given the terminology of the chapter, (2) and (4) are essentially true by definition. Hopes are then classified as either satisfiable or unsatisfiable in (6), but the distinction seems neither mysterious nor controversial. None of these opening premises appears vulnerable.

The first premise that might be challengeable is (7), since it contains a significant normative claim about what we may permissibly believe. At the root of this premise is a moral duty to pursue true beliefs instead of false beliefs.[3] What makes unsatisfiable hopes objectionable is that they are 'in a clear sense *false*' (p. 50, this volume). Generally, deceiving ourselves and others into believing false things is morally wrong. Only in dire or unusual circumstances is it morally acceptable to encourage others to have false beliefs. Perhaps one might try to object by claiming that the avoidance of despair is such a circumstance and that holding false hopes is permissible when they help to avoid it. But this defence of unsatisfiable hopes will only hold if satisfiable hopes are not available, for it is clearly preferable to take refuge in true beliefs instead of false ones. Furthermore, as stated in (8), people *can* satisfiably hope for many states of affairs. Thus, I do not think that objections to either (7) or (8) are promising. The Hope Argument still stands on firm ground up to this point.

Claim (10) is also well supported. It is unrealistic for anyone to think that he or she could live a purely egoistic existence and never encounter despair. Even the wealthiest, most fortunate among us can have our interests impeded and our personal hopes frustrated. We need at least *some* self-transcendence to avoid becoming dismayed when we encounter tragedy. The real question is *how much* we need to reliably avoid despair. Nolt's answer to that question is that we need *broad* self-transcendence—specifically, self-transcendence that extends to a significant portion of nonhuman life. First, he observes that self-transcendence is usually more enduring when it is broader in scope. This claim, expressed in (11), is bolstered by the recognition that most people cannot reasonably expect to avoid despair unless their self-transcendence extends beyond, for instance, their family and friends. Such self-transcendence is too narrow to be a reliable shield from despair; something of greater scope and duration is required.

At this juncture, we arrive at what some might identify as a weakness of the argument. The broadest self-transcendence that may seem achievable would be toward eternal goods, such as a pleasant, everlasting afterlife. The last several steps in the argument establish the following conditional claim: if there are no eternal goods, then we ought to cultivate self-transcendence toward nature. For those who are sceptical of the existence of eternal goods, this claim can be used in a basic modus ponens argument that establishes a duty to cultivate self-transcendence toward nature:

1. If there are no eternal goods, then we ought to cultivate broad self-transcendence toward nature.
2. There are no eternal goods.

3. Hence, we ought to cultivate broad self-transcendence toward nature. (1, 2)

But many believe that there are eternal goods and will be reluctant to accept the second premise of this argument. Of course, Nolt anticipates this reservation about the Hope Argument, and he argues that the actual existence of eternal goods would not necessarily undermine the imperative to cultivate broad self-transcendence toward nature (pp. 58–59, this volume).

While there is certainly room for a theistic response to the Hope Argument, I will leave that task to someone else. I consider the evidence that eternal goods exist too meagre to shoulder a substantive objection to the Hope Argument. Even if they do exist, we cannot be confident of their existence, and so we should not gamble on promoting potentially unsatisfiable hopes when satisfiable ones are clearly available.[4] Moreover, as Nolt mentions, '[h]opes for eternal goods are sometimes overwhelmed by doubt, especially in times of great loss' (p. 59, this volume). Even those raised in extremely religious environments, including those who become preachers, can eventually abandon their religious beliefs (e.g., Barker 2008; Lobdell 2009; Warraq 1995). Thus, even the most devout believers must place their hopes in some noneternal goods to assure that they can avoid despair.

The Hope Argument also appears vulnerable to a criticism that derives from its starting point. Lisa Kretz (2013) argues that the Hope Argument is an inappropriate grounding for a principle of environmental ethics because 'the *primary* motivation for adopting concerns about the goods of more-than-human others is to alleviate our own suffering in the form of despair' (930). If this is our central motivation for treating nonhuman life forms as ends in themselves, then the actions can no longer properly be described as moral actions:

> When motivation gains its foothold via 'rational self-interest' it is no longer altruism grounding such behavior, and ipso facto no longer brings to light questions of moral right and wrong so much as questions of how to best maximize self-interest. It may currently, conveniently, contingently be in our own best interest to recognize and value other's goods as goods in themselves, but there is no necessary link—it could be otherwise. I take it to be the case that actions properly described as moral require a non-contingent connection to a moral motivation. . . . Both hope and ecological philosophy premised on the primary motivation of alleviating our own suffering works against an altruistically motivated ecological ethic to protect, care-for, and respect human and non-human life forms. (Kretz 2013, 930–31)

Of course, the Hope Argument only aims to establish an account of moral valuing, but Kretz's point is that this kind of moral valuing is in itself morally suspect. If we are going to ground a moral imperative to value nonhuman life, she believes it must be based on motivations that extend beyond self-interest.

Kretz's concern is understandable but rests on an uncharitable appraisal of the Hope Argument. As Nolt acknowledges, valuing nonhuman life to enhance hope is not incompatible with other motivations for valuing it (pp. 59–60, this volume). Moreover, self-interested and altruistic reasons for valuing something frequently overlap in unproblematic ways. A basic requirement for being someone's friend is valuing that person's welfare independent of whether his or her welfare is good for you. Simultaneously, I may recognize that my life is better because I have friends and that this is a reason to cultivate new friendships and preserve the ones I have. This recognition, however, does not prevent me from making great sacrifices for my friends or genuinely wanting the best for them even in circumstances where their success might be in some way worse for me (e.g., when we both apply for the same job). Even if I begin with a self-interested motivation to make friends and sustain those friendships, my initial motivation must eventually be superseded by a nonself-interested concern for my friends. If this change in motivation does not occur, then I am not actually forming friendships: a person who is ready to abandon a 'friend' as soon as the friendship ceases to be in one's rational self-interest is not a friend at all.

On my understanding, the motivation to value nature specified by the Hope Argument follows a similar trajectory to a self-interested pursuit of friends. We can begin with a self-interested motivation to value nonhuman living organisms as ends in themselves, but if we are going to value them in the way necessary to sustain hope, we must eventually value the goods of nature as 'goods that are *not* ours' (p. 59, this volume). Otherwise, we have not actually achieved self-transcendence. As Nolt summarizes, 'Our self-interest, if it is to achieve its aim, must transcend itself, producing genuine desire for the authentic good of others, so that ultimately we desire as ends both our good and theirs' (p. 59, this volume). Thus, while we start with a self-interested motivation, the process of self-transcendence eventually leads us to possess motivations that are *not* purely self-interested. So Kretz is mistaken in her diagnosis of the Hope Argument: the pursuit of our self-interest is not the primary motivation for valuing nature once self-transcendence is achieved; it is only the motivation to *start* seeking self-transcendence. Once the process of achieving it is complete, our motivations become more complex and more other-regarding.

Kretz's criticism also fails to show that the Hope Argument is unsound. Even if one agreed with her perspective and thought that Nolt was giving a prudential argument for valuing nature instead of a moral argument for doing so, this difference would not show that any of the premises in the argument was false. If anything, it would merely shift some of the 'ought' claims in the premises to refer to prudential or rational normativity instead of moral normativity. Notwithstanding this minor change, the argument could still be sound.

Having considered some unsuccessful strategies for criticizing the Hope Argument, it might appear that the argument can survive critical scrutiny, but I believe the argument faces further difficulties that are more problematic. In the remainder of this section, I offer three criticisms of Nolt's Hope Argument that I do not think it can withstand.

## How Broad Must Self-Transcendence Be?

My first major challenge to the Hope Argument is an objection to (12). According to (12), the only way that we can reliably maintain hope through our lifetimes is through broad self-transcendence. The main problem with this premise originates from the term 'broad'. There is little doubt that a genuine egoist would have difficulty maintaining hope through his or her entire lifetime, unless perhaps he or she were to avoid the major failures and frustrations of a human life and then die abruptly in an accident at a young age (such that he or she lacks the time to contemplate and confront death). But what if we consider someone whose self-transcendence is only *a bit* broader? Nolt mentions that self-transcendence toward friends and family members is frequently not enough to avoid despair, though having a little self-transcendence is obviously better than none (pp. 51–52, this volume). That reply is reasonable, but we might now wonder whether a broader *human-centred* self-transcendence could be sufficient. Nolt's response to this suggestion is brief and important, so I reproduce it in its entirety:

> Maybe what we need, then, is a generalized hope for all humanity. But that is probably beyond the capability of many, maybe all, people. We readily develop hopes strong enough to matter to us only for what we know fairly intimately. It is difficult to even comprehend, much less hope for, the goods of billions of people. Humanity is, moreover, in many respects maddeningly foolish, short-sighted and self-destructive. We may easily despair even at the prospects of all humankind. (p. 54, this volume)

I do not consider my outlook on human nature optimistic, but I nevertheless consider this reply an unsuccessful rebuttal to the advocate of narrow, human-centred self-transcendence. In fact, this response actually reveals a larger weakness in the overall argument.

Nolt's scepticism about achieving self-transcendence toward collective humanity is unwarranted primarily because it rests on a mischaracterization of what this self-transcendence involves. If we were to actually try to wish for the good of all individual human beings, then it might indeed be impossible for most people. Typically, however, when we hope for things related to collective humanity, our hopes are more focused and more tied to our personal experience. As a graduate student in philosophy, I hope that other people, including those in the distant future, will have the opportunity to pursue graduate studies in philosophy. As a former college tennis player, I hope that tennis players of the future will also have the opportunity to play competitively at the collegiate level. As one who sees artistic value in film and literature, I hope that we will be able to preserve such works for future people to enjoy. When we achieve self-transcendence toward collective humanity, we are often achieving self-transcendence with respect to certain long-term human projects. In my examples, these projects are the continuation of academic philosophy, the maintenance of tennis as a means of recreation and competitive sport and the preservation of film and literature. Given the enormous diversity of human projects, there is a clear potential for individuals to achieve a relatively broad self-transcendence even if they only achieve it with respect to collective humanity.

Nolt also alludes to the concern that human beings are self-destructive and may undermine their own welfare over time. The irrationality and immorality of many people can be a source of great discouragement, but there are also some reasons for optimism. Historically, there have been some significant moral milestones, such as the abolition of slavery, and Steven Pinker (2011) has recently argued that we are currently living in the least violent time in human history. Even so, Nolt may be right in claiming that it is not possible *for everyone* to optimize hope by achieving self-transcendence toward humanity alone. The variation in the hope that people require to avoid despair poses a further problem for his position.

Nolt emphasizes that the duty to optimize hope is both individual and collective (e.g., p. 52 , this volume). But given the radical variability in the psychologies of individual people, we should expect great variability in the types of self-transcendence that individuals pursue and the amount of self-transcendence they need for their hope to be resilient. For many, self-transcendence toward humanity may be enough. For others, self-transcendence toward humanity and sentient nature might be enough. Others may require self-transcendence toward humanity and all of nature, and there may be certain people who will fall into despair without belief in eternal goods of some sort. Since individuals will vary so much in the amount and type of hope that they need to avoid despair, not every individual can have a duty to cultivate self-transcendence toward nature. What an individual most fundamentally ought to do is develop and sustain whatever amount of hope is necessary to avoid despair. This does not mean that we should be focused solely on our own individual hopes—parents, for instance, should consider how they might cultivate hopes in their children—but it is certainly not the case that every individual is obligated to seek self-transcendence toward nature. Even if a person accepts the basic imperative to maintain enough satisfiable aggregate hope to avoid despair, it does not follow from this commitment that they must strive for self-transcendence toward nature.

Nolt's comments about pursuing a narrower self-transcendence also reveal a deeper problem. If achieving self-transcendence toward collective humanity is impossible for many people, then achieving self-transcendence toward nature is probably also impossible for many people. We may indeed have difficulty understanding and hoping for other people's goods to be realized, especially when those people are geographically distant from us and have cultural, religious and political values far different from our own. But it is much easier to conceptualize the goods of other people than it is to conceptualize the goods of nonhumans, particularly living organisms that are not sentient. Nolt alludes to this problem when he mentions the difficulties in achieving self-transcendence toward cockroaches (p. 56, this volume), but the problem extends beyond just those creatures that we consider disgusting. In our own lives, the goods that we value are typically tied to conscious experience. We can certainly speak of the biological goods that nonsentient organisms can experience. Sunlight and rainfall are good for trees, for example.

But in ordinary human life, we do not always regard biological goods as significantly valuable unless they are tied to conscious experience in some way. This is one of the main reasons why we generally deem it permissible to remove life support from those who are in irreversible comas: without the possibility for conscious experience, their biological health is no longer of significant value. It also partly explains why we do not invest substantial medical resources in trying to extend the lives of anencephalic infants. Anencephaly is a cephalic disorder in which the majority of the brain does not develop. Infants born with this disorder usually do not live for longer than a few days and are not aggressively resuscitated because they can never become conscious.

The perspective that I have sketched above is not shared by everyone, but there is no doubt that many people share it. Some of these individuals may not perceive the goods of trees, bacteria and insects as genuinely valuable, and as a result, it may not be possible for them to achieve self-transcendence toward nonsentient life. At this juncture, we can make the following modus ponens argument against an individual duty to achieve self-transcendence toward nature:

1. Some individuals cannot achieve self-transcendence toward nonsentient nature.
2. We are not morally obligated to do the impossible.
3. Therefore, some individuals are not obligated to achieve self-transcendence toward nature. (1, 2)

I have already stated my support for the first premise of the argument. The second premise is a variation of the principle that *ought* implies *can*. According to this principle, the presence of a moral obligation presupposes that the agent with the obligation has the ability to fulfil it. Not all philosophers endorse this principle, but they agree that exceptions to the principle, if they occur at all, must be very rare: demonstrating that a particular case is an exception to this principle would be an onerous task. Moreover, Nolt appears to endorse this principle: the imperative to optimize hope includes the qualifier that we only do so 'insofar as it is reasonably possible and consistent with other obligations' (e.g., p. 53, this volume). This talk of reasonable possibility suggests that the obligation disappears when it is not reasonably possible to fulfil it. Thus, unless Nolt is prepared to make significant changes to the starting point of the Hope Argument, I believe he must accept the conclusion that some individuals are not obligated to achieve self-transcendence toward nature.

The upshot of these criticisms is that many individuals cannot have individual duties to seek or achieve self-transcendence toward nonsentient living things. Some can fulfil their duty to optimize their individual hope without achieving such broad self-transcendence. Others cannot achieve self-transcendence toward nature, and so they must optimize hope by seeking self-transcendence toward other entities. While there may be *some* individuals who ought to achieve self-transcendence toward nature, doing so will be unnecessary or impossible for many others. Thus, if there is a plausible duty to seek self-transcendence toward nonsentient life, it can only be understood as a collective duty, not as a duty that applies to each individual.

## Self-Transcendence as a Double-Edged Sword

My second objection to the Hope Argument targets (11): the claim that self-transcendence is generally more sustainable the broader it is. When we achieve self-transcendence toward someone or something, we make ourselves vulnerable to loss. The death of a friend is far more devastating than the death of a stranger. In becoming self-transcendent toward someone else, harm that befalls them can be a source of grief and frustration for oneself. In this manner, broadening self-transcendence increases the risk of experiencing such losses. Often, the reward of achieving self-transcendence is greater than the risk of loss. We know that our parents will likely die before we do, but we do not try to shield ourselves from this loss by attempting to love them less. We know that our good friends could die of fatal illnesses or accidents, but we are not so risk-averse that we remain hermits. With respect to nature, however, the risk of loss appears quite high. We do not know how many nonhuman species exist, but we know that extinctions are now occurring at an extraordinary rate. Some scientists even fear that we are in the midst of the Earth's sixth mass extinction event (Barnosky et al. 2011; Ceballos et al. 2015). A number of growing environmental problems also lurk in the foreground of the global picture: rising human population, global climate change, deforestation and resource depletion (among other problems) threaten to further deplete the planet of nonhuman life. These phenomena threaten to undermine our hopes for nature.

Nolt addresses this concern in later work. He rejects the pessimistic outlook that there is 'not much to hope for' with respect to nature because such a perspective is 'short-sighted' (Nolt 2015, 234). Life is resilient and has survived planet-wide changes of a greater severity and magnitude than we will be able to produce. Life will survive whatever we do, and even if we cause another mass extinction, life will eventually replenish itself (though the process would take millions of years). Thus, even though we are degrading nonhuman life, there is still plenty to hope for. Much of nonhuman life will persist even after we are gone.

I agree with Nolt that there remains much of the nonhuman world that can be a source of hope. The position he rejects is certainly too pessimistic. However, there is a less pessimistic position that is still a threat to the Hope Argument. Although there are certainly parts of nature that can ground satisfiable hopes, there are many other aspects of nature that will be destroyed or extinguished. If we place our hopes in these aspects of nature, we will have to endure sadness and disappointment. Thus, it is unclear whether placing hope in nature always makes us better off. After all, having a hope extinguished can be worse than never cultivating that hope in the first place. From the standpoint of avoiding despair, one may very well do better *not* to focus on nonhuman life, since human activities so frequently impede its flourishing.

Ultimately, to determine whether self-transcendence toward nature really helps us optimize hope, we need an estimation of how much hope we can gain from this self-transcendence and how much sorrow is caused by the many hopes that we will cultivate only to have them be extinguished. Attempting to make this calculation brings us to my next criticism.

## Variance in the Values of Different Hopes

My previous criticism focused on the losses that we can suffer by achieving self-transcendence toward nature and the fact that these losses can sometimes push us *closer* to despair rather than making us more resilient to it. Now let us consider the benefits associated with having these hopes. Assuming that our hopes are fulfilled, are the benefits significant? Do they make a substantial contribution to fulfilling our duty to optimize hope?

The answer depends on the kinds of creatures we are considering. The amount of hope that people attain from their pets can be extraordinary: even though there is usually some despair incurred when the animal dies, people routinely acquire additional pets afterward, more than willing to hope for a good life for their new companion. The animals with which people share these relationships are usually sentient mammals like cats and dogs. But the self-transcendence we achieve toward nonsentient nature is typically not this significant. The amount of hope attained by achieving self-transcendence toward nonsentient life seems minute compared to the hope attained by achieving self-transcendence toward sentient life and especially toward human life. This difference results from a difficulty in conceptualizing the value of life independent of conscious experience, a concern mentioned in an earlier criticism.

Even if there is some improvement in our aggregate hope that can be attained by achieving self-transcendence toward nonsentient life, it cannot be a substantial improvement. The goods of nonsentient organisms are so insignificant relative to the goods of sentient organisms and fellow human beings that they cannot make major contributions to one's aggregate hope. Even when the numbers of these organisms are extraordinary, it often makes no difference because large numbers beyond a certain size create the same impression in our mind (Broome 2004, 57–59). Whether we are speaking of the goods of ten trillion bacteria or ten quadrillion bacteria will not make a difference in how much our aggregate hope is affected.

Perhaps nonsentient life could still make a large contribution to aggregate hope if we reframe our hope in terms of species instead of individual organisms. There are so many nonsentient species that hoping for the long-term survival of all these species can bolster aggregate hope significantly even if the hope contributed by each individual species is very small. Even this claim is doubtful, however. I suspect most parents will have a greater improvement made to their aggregate hope via self-transcendence toward their child than they could ever achieve through self-transcendence toward any number of nonsentient species. In other words, self-transcendence toward *one* human being can eclipse the aggregate hope fostered through self-transcendence toward *all* of nonsentient nature. Moreover, the gap between the quantity of aggregate hope cultivated by self-transcendence toward nature and the quantity of aggregate hope cultivated by self-transcendence toward collective humanity (rather than just one person) is *far* larger. The upshot of these observations is that self-transcendence toward nonsentient nature may not make a significant contribution toward increasing aggregate hope, particularly when compared to the human-oriented sources of self-transcendence at our disposal.

## AN ALTERNATIVE VIEW

If my criticisms in the prior section are correct, then the Hope Argument is unsuccessful: optimizing hope does not require achieving broad self-transcendence toward nature. Nevertheless, I believe that the Hope Argument is correct in some important respects. In this final section, I take stock of what my prior analysis has shown and suggest an alternative view regarding self-transcendence toward the nonhuman world, one that is nonanthropocentric in character but not as broad in scope as the position that Nolt defends.

The Hope Argument gets at least two crucial details right. First, everyone needs *some* self-transcendence to reliably avoid despair. Second, self-transcendence toward human beings alone will not be enough for everyone to optimize hope. Given the uncertainty about the existence of eternal goods, it follows that some self-transcendence toward nonhuman life is required. However, my criticisms indicate that seeking self-transcendence toward nonsentient nature is unlikely to make much progress in optimizing hope. Achieving self-transcendence makes us vulnerable to significant losses that can actually bring us closer to despair. We may still be willing to risk these losses if the potential gain in our aggregate hope is high, but self-transcendence toward nonsentient life cannot improve aggregate hope significantly. Thus, rather than striving to achieve self-transcendence toward both sentient and nonsentient nature, I propose that we limit our self-transcendence to sentient nature. Since we have a far better understanding of the goods of sentient organisms and recognize these goods as deeply valuable (particularly in the case of certain domesticated animals), the potential improvement to aggregate hope is great enough to justify the gamble of making ourselves vulnerable to the sorrow associated with extinguished hopes.

I also believe it is possible to ground an *individual* duty to achieve self-transcendence toward sentient life. Although I claimed that the Hope Argument can only ground a collective duty to achieve self-transcendence toward nature, there exists an independent argument that can ground a corresponding individual duty with regard to sentient nature. The argument's starting point is the observation that we do not consider infants, children or the cognitively impaired to lack moral standing. We ordinarily believe that their welfare matters with respect to the moral decisions that we make, and this conviction is deeply shared. How do we explain this phenomenon? We cannot appeal to rationality because these individuals are not rational in any plausible definition of the term: some of them have never been rational and will not ever be rational. We also cannot appeal to their humanity because doing so would be a blatant manifestation of speciesism—'a prejudice or attitude or bias in favor of the interests of members of one's own species and against those of members of other species' (Singer 2002, 6). Appealing to biological humanity to ground a difference in moral status would be analogous to making such a distinction based on one's race or sex.

This argument, sometimes labelled the Problem of Marginal Cases, is a well-established challenge to the claim that the moral community is limited to human beings. Both Peter Singer (2009) and Tom Regan (1989), perhaps the most well-known animal ethicists, have used it to cast doubt on the notion that human beings are the only creatures with moral standing, and many other philosophers have also examined the argument in detail (e.g., Dombrowski 1997; Norcross 2004). Because our considered judgments about the moral status of infants, children and the cognitively impaired are so strong, the problem of marginal cases forces us to consider whether there is something beyond being human or being rational agents that is sufficient for moral standing. Sentience, understood here as shorthand for the capacity to experience pleasure and pain, is a plausible candidate. Sometimes, of course, the explanation for why it is wrong to cause someone pain can be traced back to his or her rationality: perhaps causing him or her pain disrespected that person or failed to properly account for his or her considered preferences. In other cases, however, this kind of explanation cannot be given. If I pour boiling water on the hand of a cognitively disabled individual for my own amusement, I certainly cannot explain the wrongness of my action in an appeal to the harmed party's rationality.[5] The most straightforward way to explain the wrongness is to simply claim that suffering is morally bad, regardless of whether those who suffer are rational.

Alternatively, one could appeal to the virtues and stress that such an action cultivates the vice of callousness. But this manoeuvre only raises a further question: just why is this action callous? Presumably, it is callous because it demonstrates a deplorable indifference to suffering. Such an explanation is reasonable, but it only furthers the claim that sentience is a sufficient condition for moral standing because it shows that the suffering of nonrational human beings is morally significant.

Another alternative explanation would be to suggest that acting with this indifference toward suffering would likely manifest in one's dealings with rational human beings, which would be morally objectionable. The problem with this proposal is that it grounds the wrongdoing entirely on an empirical claim about the likelihood of my future behaviour. This empirical claim may well be false: I might be very good at moral compartmentalization and make sharp, consistent distinctions between how I treat rational human beings and nonrational human beings. In that case, my action will not be wrong because I will not in fact be at greater risk of treating rational human beings worse in the future. This explanation is also dubious because it tries to explain the wrongdoing of *actual* harm done to someone by considering the *possible* harm that I might do to others in the future. In this manner, the explanation ignores the most salient moral feature of the case—the fact that I caused someone to suffer unnecessarily.

The only promising solution to the Problem of Marginal Cases is to acknowledge that being sentient is sufficient for moral standing. If we accept this claim, then we can delineate at least two distinct criteria for moral standing: rationality and sentience.[6] Sentient creatures that lack rationality still warrant direct moral standing because they can be harmed by our actions. They can suffer. Thus, we must take their welfare into account when we make moral decisions. This does not necessarily mean that we should treat nonrational human beings in the same ways as animals: a variety of other factors—interpersonal relationships, social and cultural values, the symbolic significance of how disabled individuals are treated and many more—complicate the moral picture. The point is that sentient animals cannot be disregarded in our moral decision-making: the ways in which our actions affect their welfare must be taken into account when we act. Acting in this manner is simply a matter of moral consistency. If sentient, nonrational humans have direct moral standing, then sentient, nonrational animals must also have direct moral standing. Thus, just as we regard the welfare of sentient, nonrational humans as an end, we must regard the welfare of sentient animals as an end. In this manner, the imperative to be self-transcendent toward merely sentient human beings grounds an imperative to be self-transcendent toward sentient nature.

Taken in unison, the Problem of Marginal Cases and the Hope Argument direct us toward the following conclusion: we ought to cultivate self-transcendence toward sentient nature. The Hope Argument can only posit this imperative as a collective duty because some will be able to optimize their individual hopes through self-transcendence toward humanity alone, but its companion argument does not have this limitation. Unless we are willing to abandon our judgments about the moral status of infants, children and those with cognitive impairments, all of us must regard sentient creatures—including nonhumans—as having direct moral standing. By bridging the two arguments, we can conclude that we (individually and collectively) ought to cultivate self-transcendence toward sentient life forms. Adhering to this imperative is not only crucial to maintaining moral consistency but also our best means of optimizing hope.

## NOTES

1. Not all biocentrists may find Nolt's argument satisfactory, however, because it does not advocate or require self-transcendence toward *all* living things: 'We need not be self-transcendent toward all of [nature]; but given the current threats to nature, we will best preserve satisfiable hope by being self-transcendent toward a good bit of it' (p. 44, this volume).

2. This qualifier—'insofar as this is reasonably possible and consistent with other obligations'—is implied throughout the argument, though I omit it from the sub-conclusions for

brevity. I consider this restriction a fundamental aspect of any moral principle, one that goes hand-in-hand with the reality that all or *virtually* all moral principles are prima facie principles rather than absolute ones.

3. We could also speak of an *epistemic* duty to pursue true beliefs rather than false ones, but Nolt's concern is only with moral duties. If (7) were to identify an epistemic duty, then the various 'ought' claims that follow would become confusing because the relationship between moral duties and epistemic duties is a topic of great controversy. For a survey of this issue, see Haack (1997). Fortunately, since Nolt is only speaking of moral duties, this issue does not need to be resolved to assess the Hope Argument.

4. Some might object that there is in fact sufficient evidence for the existence of eternal goods. A comprehensive reply to this objection would take us far afield of my main task in this chapter, but I will offer a brief response based on the presence of disagreeing epistemic peers. (For an overview of the epistemic significance of disagreement, see Christensen 2009.) There are many atheists and others who are sceptical of eternal goods' existence, and some of these individuals are just as intelligent and well-educated as those who believe that eternal goods exist. In fact, a recent study revealed that more than 70 percent of philosophers 'accept' or 'lean toward' atheism (Bourget & Chalmers 2014), which suggests that a scepticism about the existence of eternal goods is fairly common among those who have critically reflected on whether they exist. The presence of these sceptics about eternal goods provides some evidence that belief in eternal goods is unjustified. Thus, even those who ultimately believe in eternal goods should have reservations about the existence of such goods.

5. Also assume that this act is done privately and thereby avoids the complications associated with how other (rational) individuals might react to it.

6. Jon Garthoff (2011) has developed this two-locus view of moral standing in far greater detail than I can here. Although he develops this approach within a Kantian framework, this account of moral status is compatible with many ethical theories.

*Chapter Six*

# Hope, Self-Transcendence and Biocentrism

John Nolt

In 'Hope, Self-Transcendence and Environmental Ethics', I offered a prescription for despair—a malady that I see in many of my students. I teach an environmental ethics course that begins with a review of the findings of environmental science. These are, of course, worrisome. Some students resent my efforts to describe the breadth and magnitude of the problems; they would rather not be reminded. Some are resolutely apathetic. A few cope by denial. Nearly all—even the brightest, most engaged and most optimistic—wrestle in one way or another with despair.

Many people resist despair by cultivating otherworldly hopes. But, as I argued in that chapter, many, if not all, such hopes are unrealistic, not all are sustainable and for deep-rooted sceptics none are possible. Worse, such hopes may foster neglect of earthly life. For these reasons, I advocated an alternative (or, for those who rely on otherworldly hopes, supplementary) prescription for avoiding despair: cultivate self-transcendence toward a diversity of living beings, both human and nonhuman, sentient and nonsentient. This entails coming to value for their own sakes *earthly* goods more diverse and enduring than any human goods. It therefore makes possible earthly hopes that are realistic, temporally deep and highly diverse—hence sustainable.

The present chapter elaborates these ideas in three ways. First, I confess some second thoughts since writing that article. Next, I reply to objections raised by Trevor Hedberg in 'Optimizing Hope: A Response to Nolt' (chapter 5 of this book). Finally, I sketch a line of reasoning from the thought that we ought to cultivate broad self-transcendence to a biocentric ethic.

## SECOND THOUGHTS

Since writing 'Hope, Self-Transcendence and Environmental Ethics', I have had five 'second' thoughts.

First, since we can realistically hope only for goods that stand a fair chance of being realized, at least in the long term, I wish I had better emphasized life's resilience. Many people unreflectively assume that we are on the verge of destroying the Earth, nature or life itself. Not so. We can't destroy or sterilize the planet. Death Stars exist only in Star Wars movies. Life has, moreover, persisted for at least 3.5 billion years, sometimes under crushingly adverse conditions. It teems at boiling hot hydrothermal vents in the depths of the oceans. It has survived snowball Earth (or something like it) under kilometres of ice. It has adapted to radical alterations of the composition of the atmosphere and to global average temperatures hotter than any we are likely to produce. It has recovered from at least five mass extinctions.

John Muir celebrated life's resilience in his notebooks:

> There need be no lasting sorrow for the death of any of Nature's creations, because for every death there is always born a corresponding life. And what life shall follow the death of the glacier, what creation shall come to that sea bottom on whose cold burnished rocks not a moss or dulse ever grew! In smooth hollows crystal lakes will live, to sandy beds sedges will come. Pines and firs will feather the moraines, advancing like and army . . . and instead of a robe of white ice will be a robe of yellow light upon the new Edens of the Sierra! (Muir 1979, 168–69)

Muir should not be read too literally, of course. There is no precise correspondence between a death and a new life. Yet in the big picture he's right. Biodiversity tends to increase. After every die-off life rebounds, usually with greater richness than before—though for mass extinctions, this can take millions of years (Kirchner and Weil 2000).

Second, I wish I had been more explicit about the problem that motivated me. As an environmental activist for nearly three decades, I have seen few victories and many defeats. I've studied climate science and scoured reports of biodiversity loss and population increase. I know the trends. I wrote the chapter not merely as an academic exercise, but as a way of thinking through the possibilities of realistic hope for my students and myself. What motivated me was the threat of despair that haunts anyone who understands the historical trends and the findings of environmental science.

Third, I should have made it clearer that I was not trying to reduce biocentric valuing to anthropocentric self-interest. I said explicitly that humanity's need for hope is *not* the only possible reason for cultivation of self-transcendence. But it was the only reason that I gave. This, apparently, led some readers to infer that I think we should value nonhuman life *only* to stave off despair. Not so. I saw that chapter as a step in a dialectic between me, a biocentrist and readers, many of whom I supposed to be anthropocentrists. So as not to beg the question against such readers, I limited my assumptions to premises that I thought anthropocentrists could accept. But I did not mean to suggest that the *only* reason to value the well-being of a wide diversity of living things is the human need for hope. There are, on the contrary, multiple reasons and multiple paths to biocentric valuing. Many naturalists, for example, have come to biocentric valuing through extensive youthful exploration of and delight in wild places (see Wilson 1994 for a great example). This path requires no thought at all of avoiding despair—though it may yield later in life the unexpected bonus of sustainable hope. But I could not assume such experiences on the part of readers.

Fourth, I should have said clearly that despair is not depression. At least one reader—prompted, perhaps, by my remark about rampant use of antidepressants—interpreted my argument as a moral diatribe against depression. But that was far from my intent. Despair and depression are distinct. Depression is an emotional state that generally arises involuntarily. Despair is absence of hope. It need not be accompanied by depression. It may be angry, or steely and resolute. Though almost always coloured by emotional pain, it is not just emotion, but largely a constellation of cognition and desire. Whether we despair in a given situation therefore depends on what we believe and what we value. Over these things we have some control. We may thus be able to prevent or relieve despair by modifying our beliefs or values.[1] Rarely, if ever, would this work for depression. To say it once again, I argued for broad self-transcendent valuing, as a way to prevent or mitigate *despair*. I did not advocate the foolish and counter-productive policy of blaming the depressed for their depression.

My fifth and final 'second' thought is that I should not have expressed the article's prescriptions as duties. In many idiolects, 'duty' connotes a binding command of legitimate authority (God, the Moral Law, social customs or contracts, legal norms or the like). The imperative of self-transcendence is not such a binding command. It is, rather, a prescription for fulfilling our individual and collective need for sustainable hope. Its aim is moral: better human lives and ultimately a better world for all life. But we need not think of morality as commanded by authority or reified in duties.[2] The reader can excise the unintended connotations from the paper by replacing phrases like 'we have a duty to *A*', where *A* is an action, with phrases like 'we should *A*', and replacing phrases like 'we have a duty to *i*', where *i* is an individual, with phrases like 'we should with regard to *i*'.

## HEDBERG'S OBJECTIONS

In 'Optimizing Hope' (chapter 5 of this book), Trevor Hedberg presents a clear, concise and accurate summary of the central argument of 'Hope, Self-Transcendence and Environmental Ethics'. On the basis of this summary, he raises three objections against that argument: (1) since people can optimize hope in diverse ways, a general duty to optimize hope does not entail for each person a duty to optimize it by self-transcendence toward nature in particular; (2) instead of enhancing our ability to optimize hope, self-transcendence toward nature may make us vulnerable to losses that crush hope; and (3) since the hope derived from self-transcendence toward nonsentient entities is minute compared to that derived from self-transcendence toward sentient beings, the former may not contribute significantly to the optimization hope. I'll consider each objection in turn.

### Self Transcendence toward What?

Hedberg's first objection is that the general duty to optimize hope need not entail, in particular, a duty to seek self-transcendence toward a wide range of nature's living things. He agrees that we need nonegoistic hope, and hence self-transcendence, but he maintains that for many people self-transcendence toward humanity suffices:

> As a graduate student in philosophy, I hope that other people, including those in the distant future, will have the opportunity to pursue graduate studies in philosophy. As a former college tennis player, I hope that tennis players of the future will also have the opportunity to play competitively at the collegiate level. As one who sees artistic value in film and literature, I hope that we will be able to preserve such works for future people to enjoy. (xxx, this volume)**[AQ1]**

I do not doubt that self-transcendence toward family, friends, descendants and people who share our interests and projects can be adequate for sustaining hope—especially during stable, peaceful, prosperous times, when the people we love have a decent chance of living well. For some of us, these are such times.

But there is reason to think that they may not last. The instruments of stability, peace and prosperity have been powered mostly by fossil fuels. Either we will persist in burning vast quantities of fossil fuels and accelerating climate change (not much hope there) or we will cut back deeply and quickly, slowing climate change, but risking global economic and political instability. The military establishments of many nations are already preparing for the anticipated violence. In both scenarios, population growth will persist, building further pressure for conflict. We may still hope to retain peace and prosperity by a quick transition to renewables and nuclear energy, but as world-wide carbon emissions continue to rise, that hope diminishes. Much of history has been neither stable nor peaceful nor prosperous. Posterity is not immune to similar misfortune.

During tumultuous times, people can have little realistic hope for family, friends, descendants, institutions or intergenerational projects. To avoid despair, they commonly resort to hope for some better world hereafter. But there are intriguing exceptions. Some Native Americas during the time of the great genocide, for example, found poignant hope in self-transcendence toward earthly life. Most, it is true, believed in an afterlife, but not in a better one. They did not put their hopes there. Their abiding love and hope was for those whom they regarded as relatives—the Earth and all its living creatures.

John G. Neihardt reports that long after the defeat of his people and the destruction of their way of life, Black Elk, the Ogalala Sioux holy man, spoke this invocation:

> Hear me, four quarters of the world—a relative I am! Give me strength to walk the soft earth, a relative of all that is! Give me the eyes to see and the strength to understand, that I may be like you. With your power only can I face the winds.
> Great Spirit, Great Spirit, my grandfather, all over the earth the faces of living things are all alike. With tenderness these come up out of the ground. Look upon these faces of children without number and with children in their arms that they may face the winds and walk the good road to the day of quiet. (Black Elk and Neihardt 1979, 4–5)

Black Elk's hopes were for nonhuman, as well as human, life.

He did not turn to such broad self-transcendence in desperation, after other hopes had failed. He imbibed it, rather, with his culture. The Sioux valued the welfare of nonhuman beings as ends. They also sought solace in the permanence of the Earth. Black Elk's friend, Iron Hawk, recalled some advice given to him when he was a fourteen-year-old warrior preparing to defend his camp against onrushing federal troops: 'A man by the name of Little Bear rode up to me on a pinto horse, and he had a very pretty saddle blanket. He said, "take courage, boy! The earth is all that lasts!"' (Black Elk and Neihardt 1979, 92).

Though their culture was devastated, some of the Sioux survived. Similarly, humans of today's global culture—as numerous, intelligent and tenacious as we are—are likely to survive, at least for the next few centuries, come what may. But, as with the Sioux, our projects and culture may *not* survive in ways that fulfil our hopes. Hence if our hopes are to be realistic, they may need to be broader than those to which we are accustomed.

Hedberg rightly notes that the capacity for self-transcendence and hence for broad hope varies widely among individuals. He infers that, although individuals should cultivate enough self-transcendence to make their hopes resilient, there is no need to cultivate more. But how much is sufficient is unknown, since it depends on the course of future events. If we knew the future well enough, we could, perhaps, modulate our self-transcendence in the way Hedberg suggests. But that isn't practical. In an unstable world, long-term realistic earthly hope is likely to be sustainable only to the degree that it is diverse, broad and open to multiple possibilities—and this requirement may not vary so widely among individuals.

Finally, Hedberg argues that if (as I have conceded) achieving self-transcendence toward collective humanity is for many people impossible, then achieving self-transcendence toward a diversity of living things must be impossible for many more. My prescription, in other words, violates the requirement that 'ought' implies 'can.' It demands the impossible.

I agree that self-transcendence toward humanity *in the abstract* is impossible for most of us. And I think the same of self-transcendence toward all life *in the abstract*. But we have no need to value such abstractions. It is the diversity of self-transcendence, not its abstract universality, that makes hope sustainable.

I also agree with Hedberg that highly diverse self-transcendence is not possible for most people *on the spur of the moment*. It must be cultivated through extensive personal experience with and appreciation of nonhuman life. Yet since many of the Sioux achieved it, it is not impossible.

## Self-Transcendence and Vulnerability to Loss

Hedberg's second objection is that diverse self-transcendence toward nature's living things may make us vulnerable to great sorrow—especially given current and projected rates of extinction and biodiversity loss. 'The death of a friend', he reminds us, 'is far more devastating than the death of a stranger' (xxx, this volume).**[AQ1]** He thus concludes that we might 'do better *not* to focus on nonhuman life, since human activities so frequently impede its flourishing' (xxx, this volume).**[AQ1]**

It is true that loving much makes us more vulnerable to much sorrow. But that doesn't mean that it is better not to love. For one who is broadly self-transcendent, even great loss and sorrow need not breed despair. Much of the natural world is already lost, and there will be greater losses of nonhuman life in the foreseeable future. But life is so rich, resilient and diverse that hope for new life remains, and will remain, possible.

The person who values more narrowly is in fact the more vulnerable to despair. When a person values broadly and hopes for many goods, the proportion of hope invested in each is small. If one hope fails, the loss is relatively less than if that person had had fewer hopes. A person whose one and only hope fails is thereby in despair. But if he or she had had the same hope and also others, he or she need not be in despair. The contrary view—that when a person has many hopes, the failure of one is as devastating as if it were that person's only hope—is not credible.

## The Value of Hope for the Welfare of Nonsentient Beings

Hedberg's third objection is that the value of the hope attained by self-transcendence toward *nonsentient* life is miniscule compared to the hope attained by self-transcendence toward sentient life—especially sentient *human* life; hence its contribution to aggregate hope may be negligible. 'The goods of nonsentient organisms', he says, 'are so insignificant relative to the goods of sentient organisms and fellow human beings that they cannot make major contributions to one's aggregate hope' (xxx, this volume).**[AQ1]**

I agree that the goods of nonsentient organisms are comparatively miniscule (Nolt 2015, 178–80). But since Hedberg infers that the value of hope for a nonsentient being is insignificant compared to the value of hope for a sentient being, he must be assuming that the significance of a hope is proportional to the value of its object.

I see no reason to accept this assumption. The significance of hope for us lies primarily in fulfilment of our need for hope. If a person is, like E. O. Wilson (see 1994 and 2010), an aficionado of ants—which, presumably, are nonsentient[3]—then his or her hopes for ants may matter much, even though there are many kinds of sentient beings for which he or she has only miniscule hopes or no hopes at all. (Ants, incidentally, are among the most resilient of Earth's creatures; broad realistic hopes for them are seldom vain.)

Yet even if the value of hope for a nonsentient organism—a tree, say—were miniscule relative to the value of hope for a human being, it would not follow that the value of hope for the continued life of trees is miniscule relative to the value of hope for humans. There are far more trees than humans and vastly more kinds of nonsentient beings than sentient ones. Though a single tree may be little to hope for, the perpetuation of a great forest may be much.

Yet even hope for a great forest is unlikely to exceed our hopes for the people we love. As Hedberg notes, 'self-transcendence toward *one* human being can eclipse the aggregate hope fostered through self-transcendence toward *all* of nonsentient nature' (xxx, this volume).**[AQ1]** It is true that many people value their partners or children above all other living things. And many have no hopes at all for nonsentient life. Still, because partners and children may die or our relationships with them may dissolve, resting our hopes in them alone makes us vulnerable to despair.

We need not—perhaps should not—value nonsentient or nonhuman beings as highly as we value the people we love. Still, hopes for nonsentient and nonhuman life can add significantly to our aggregate hope, make it more sustainable and help to avert despair even when other hopes fail.

## VALUING, HOPING AND ACTING

In 'Hope, Self-Transcendence and Environmental Ethics', I acknowledged that our capacity for self-transcendence is limited. Hence I advocated self-transcendence only toward a wide diversity of living things, not toward all. For this reason, Hedberg characterizes the view of that chapter as a 'partial biocentrism' (xxx, this volume).**[AQ1]** Partial it is, but, to be clear, it is not a biocentric *ethic*. It was intended, rather, as a theory of environmental valuing. It offers prescriptions for *what to value*, but not (with the exception of some brief remarks at the end) for *what to do*. To move from the former to the latter—that is, from a theory of environmental valuing to a biocentric ethic—requires further reasoning. I'll conclude this chapter by sketching such a move.

It starts with the principle (for which I have already argued) that we ought to cultivate self-transcendence toward a wide diversity of living things—that is, that we ought to value for their own sake the goods of such beings. Add to this the commonplace idea that integrity is a virtue—that is, that our actions ought to reflect our values. It then follows that we should not treat these beings as mere objects or means, but rather let our actions be shaped by concern for their good. We ought, in the lingo of moral theorists, to *respect* them. Call this conclusion—the conclusion that we ought to respect a wide diversity of living beings—*the biocentric respect principle*. The biocentric respect principle goes beyond a theory of environmental valuing into biocentric ethics, for it prescribes (though only in a vague and abstract way) what we ought to do, not merely what we should value.

To summarize: the biocentric respect principle follows from an imperative of integrity together with the premise that we ought to cultivate self-transcendence toward a wide diversity of living beings, which in turn was derived from the human need to optimize hope.

Lisa Kretz (2013) objects to this reasoning on the ground that it motivates the biocentric respect principle, not altruistically, but by rational self-interest. It implies, she thinks, that we should respect living beings primarily in order to 'alleviate our own suffering in the form of despair' (Kretz 2013, 930). She maintains, however, that properly *moral* actions must be motivated altruistically and not by self-interest.

I disagree with Kretz in two particulars. First, my argument did not appeal to rational self-interest, but to a general human need for hope. The relevant assumption *was* altruistic: that there is a general responsibility to satisfy this need for hope, not only for ourselves, but for other people. I supposed, as I noted above, that many readers would be anthropocentrists, but I made no special attempt to address rational egoists.[4]

Second, I doubt Kretz's assumption that properly moral actions must be motivated altruistically. Since connection between motivation and action is quite fluid, altruism is seldom, if ever, pure. What begins as self-interest can over time evolve toward altruism. But it also sometimes reverts back toward egoism again or oscillates between the two. We may, even as egoists, become widely self-transcendent to enhance our own hope, yet that self-transcendence may eventually become habitual and largely altruistic. Hedberg (xxx, this volume)**[AQ1]** illustrates this point nicely in his discussion of friendship. The moral praiseworthiness of an action may well increase in proportion to the altruism of its motive. But respect for other living things is a step in the right direction, even if selfishly motivated.

Now, however, a further objection arises. The biocentric respect principle recommends respect only for some, not all, living beings. It therefore really is partial in Hedberg's sense. Worse, it is partial in the sense opposed to 'impartial'. For if we merely respect those living things toward which we have cultivated self-transcendence, selection of the objects of respect is left to personal predilection. Such an ethic is plainly too subjective. Imagine, for example, trying to weave an ethical public policy from such partial threads.

Yet we can't respect *all* living things. Life persists only by suppressing other life. We must kill disease organisms. We must weed our gardens. To think we can nevertheless respect the very germs or weeds we kill (i.e., let concern for their good shape our actions) would be self-congratulatory nonsense.

It follows that any practical biocentric respect principle must be partial, at least in the sense that it does not demand respect for *all* living things.[5] (This, by the way, is true even if we restrict the scope of the respect principle to sentient beings. For how could we respect the mice, moles and voles that are crushed when we plow our fields, or the birds that are clobbered by our windmills?) But a partial respect principle cannot constitute a satisfactory biocentric ethic.

My reasoning has now taken a peculiar turn. I started by advocating self-transcendence toward a wide diversity of living beings. Integrity then demanded respect for those beings. But now I have concluded that such respect must be partial and therefore does not constitute a satisfactory ethic. What gives?

Nothing. This peculiar turn does not impugn any conclusion so far. It provides no reason, for example, not to respect widely, even though we can't respect universally. What it shows is merely that we do not yet have all the ingredients for a satisfactory biocentric ethic. The chief missing ingredient, in my view, is a nonsubjective, nonarbitrary principle for determining when we should and when we need not respect a living thing—that is, when ought to let concern for a being's good shape our actions and when we may neglect its good in favour of the goods of other living things, including ourselves.

The key to such a nonsubjective, nonarbitrary choice principle is the realization that some goods are objectively greater, either in quantity or in quality, than others. The life of an elephant is a qualitatively richer in goods than the life of an ant. The continued lives of three elephants are a quantitatively greater good than the continued life of just one of them. Thus, there are objective reasons to prioritize protection of the lives of elephants over protection of the lives of ants and, similarly, to prioritize protection of the lives of three elephants over protection of the life of just one. Such a prioritization ranks actions according to the objective goodness of their consequences. To choose action in accord with it is to adopt a biocentric consequentialist ethic.

Consequentialist ethics are, of course, subject to many objections and pitfalls. But this is not the place to sort all these difficulties out. I have dealt with many of them in Nolt (2015, chapters 6 and 7). I'll close by mentioning just one point in favour of such a theory—a point of central relevance to this chapter. Biocentric consequentialism seeks to shape the world so that objective good predominates over objective bad—so that life flourishes rather than declines. Should it succeed, it would preserve and perhaps magnify possibilities for realistic earthly self-transcendence—and hence for sustainable earthly hope.

## NOTES

1. This is explained in the section titled 'How Self-Transcendence Sustains Satisfiable Hope' in 'Hope, Self-Transcendence and Environmental Ethics' (chapter 4 in this volume).
2. For more on how I think it should be understood, see chapter 2 of Nolt (2015).
3. Wilson would agree, as he told me in conversation. This example was inspired by a reading of Wilson (2010), whose middle section is a fascinating epic set in the world of the ants.
4. I regard rational egoism as amoral and have elsewhere sought to refute anthropocentrism by demonstrating its similarities to rational egoism (Nolt 2013).
5. Thus we must reject respect principles of the sort advocated by Taylor (1986).

*Part III*

# Hope and Virtue

*Chapter Seven*

# Hope in the Age of the Anthropocene

Brian Treanor

## ENVIRONMENTAL DESPAIR

Today we are faced with all the traditional reasons to despair: poverty, loneliness, loss, tragedy, death and the like.[1] And, for many, this despair is exacerbated by the modern disenchantment of the world, a postmodern suspicion regarding grand narratives (especially those speculating about transcendence) or both. The news of the day sounds a relentless drumbeat of woe. As I write these words on a rainy morning in Southern California—itself a depressing reminder of the apocalyptic drought my state is suffering, and the anthropogenic climate change that is likely to make such droughts more common and more severe—the headlines include the ongoing brutality of the 'Islamic State' in Iraq and Syria; increasing tensions between Russia and the West, including frightening near-misses involving unregistered military aircraft; the still-smoldering catastrophe of Ebola Zaire in West Africa (and parallel, though much less publicized, stories of MERS and H5N1, either of which, in a pandemic form, would give us all personal insight into the experience of medieval villagers facing the Black Death); epidemics of sexual assault and rape in both India and on North American college campuses; and stories about the widening economic inequality, and the possibility of another global economic downturn.

Dwelling too long on any one of these topics might induce despair in those inclined to melancholy. However, heaped on top of these headline issues—or perhaps dwelling in the background—environmentally aware people discern potential catastrophes the depth, breadth and complexity of which are cause for despair even in those with generally more sanguine natures. As a result, environmental philosophers—as well as activists, scientists and not a few economists—have tended toward extreme pessimism in their evaluation of the environmental realities at the dawn of the twenty-first century. And with good reason. The prospects for global climate change alone—given the *wholly* inadequate response by political and economic decision-makers—are enough to drive a well-informed person to drink.

True, every generation feels, to some degree, that it is living at the end of the world. One might argue that the sack of Rome by Brennus and the Senones in 387 BC,[2] the Black Death of the 1300s, the Great War (which, to our disappointment, did not 'end all wars'), the Shoah, the Cold War and the Cuban Missile Crisis all felt like the coming of the Apocalypse to those affected. But climate change—the watermark, so to speak, of the age we have come to call the 'Anthropocene'—is a threat of a fundamentally different order; it presents us with a challenge, and a despair, that is unique.[3]

First, unlike previous threats, climate change accelerates not when things go wrong, but when things go right. Unlike, say, the pollution of a river, which occurs when someone dumps effluent rather than disposing of it properly, climate change accelerates as part of a kind of 'malign demographic transition'.[4] When people become more affluent, they consume more; because every single aspect of modern life is utterly dependent on the extravagant use of fossil fuel, more consumption leads to more climate change. When people consume, even frugally, when they commute to work, eat food, turn on the tap, read at night and basically anything else one does in modern life, they contribute to climate change (McKibben 2007, 5–45 passim). Climate change confronts us with a problem that is not the result of ill will or ignorance. We know what we are doing, and we often don't want to do it, but continue on our way.[5]

Second, unlike a war or even a plague, climate change is a slowly unfolding catastrophe. Like watching a child grow, it is often difficult to see the incremental changes; that is, until one is confronted with a sudden growth spurt or, in the case of climate change, a tipping point. In any case, it is enormously challenging to get people to radically change behaviour on the basis of incremental changes, the full effects of which will not be apparent for twenty, fifty, or a hundred years.

Third, climate change is an example of a 'commons problem', in which responsibility is distributed (Hardin 1968). In a commons, individual virtue, such as lower carbon emissions, is utterly insignificant because all the other members of the commons are incentivized to cheat and free-ride. Even if we convince a few members of the commons to behave reasonably, others will persist in their behaviour, perpetuating climate change, which is not tied to a single individual, nation or culture. In a global commons such as the atmosphere and climate system, we need to get everyone on board with a solution or there will be no solution.

Finally, climate change raises the spectre of not only physical apocalypse, but also spiritual or existential crisis. It is not that we've destroyed the world in a fit of anger and hatred, as might have happened with a nuclear holocaust, but that we've neglected it, abandoned it, betrayed it in a slow-burning, short-sighted orgy of greed. We are destroying the world because we have turned our back to it, denying our earthy origins and our inextricable link to a healthy ecosystem (*humus*/earth, *humus*/human). One might argue that, even if we were to survive the next century by transitioning toward some synthetic and hyper-technological way of being, we would preserve our lives at the cost of our humanity. Something important would be lost in such a total embrace of the 'technological' over the 'natural'.[6] Indeed, some argue that the latter effects have already been felt. Climate change has, by some accounts, initiated a break from nature, even the 'end of nature'.

I confess that I harbour a deeply pessimistic streak with respect to many things, but climate change takes the cake. When I look at the totality of our situation—our utter dependence on a carbon-intensive economy, the stranglehold that fossil fuel lobbies have on political systems, the ignorance and intransigence of a nonnegligible portion of the population regarding this issue, widespread indolence and indifference, our general reluctance to make sacrifices, distributed responsibility leading to a lack of accountability, the slow speed of a catastrophe that cannot be easily captured in 'sound-bites' or thirty-second 'news' stories, the fact that tipping points and other nonlinear threshold events mean that late action could be as bad as no action at all and so on—I see no good reason to wager that we will be able to avoid runaway climate change.[7] Never has our flirtation with doom been so clearly supported by science, so global in scope and so clearly identifiable and understandable, while simultaneously seeming so inexorable, so completely inescapable. We are caught in an exceedingly complex, slowly unfolding, globally inclusive *tragedy*, in the sense Garrett Hardin uses that term: condemned, not by the gods, but by the 'remorseless working' of the free market in an age of fossil fuel dependence.[8] The tragic chorus—in this case made up of environmental scientists and activists—warns us of our approaching doom, and yet we march on with the 'inevitableness of destiny', resigned to the 'futility of escape' (Hardin 1968). Welcome to the Anthropocene. Eat, drink and be merry, for tomorrow we will die—and, perhaps, a good deal of the planet with us.

Nevertheless, I intend to argue that hope, properly understood, is the proper response to the angst of the Anthropocene. Although in previous work I have argued for simplicity and political engagement as keystone environmental virtues, hope is, in its own way, essential for the full realization of these, and other, dispositions. Hope is an essential virtue for our time.

## THE NATURE OF HOPE

How, in the face of such calamities, so many reasons to wager that things will turn out poorly, can I suggest that hope is an appropriate response to our situation? In such circumstances, isn't hope a flight from reality? The last refuge of minds too weak to accept the world as it is: chaotic, threatening and utterly indifferent to our desires? A rickety crutch for those who believe—not only without justification, but generally without even an *attempt* at justification—that 'everything will be fine, we'll figure something out'? Certainly some forms of 'hope' are open to such criticism. But a philosophically robust notion of hope reveals it to be distinct from, and in some cases at odds with, other dispositions or attitudes with which it is commonly confused: irrational optimism, delusional wish fulfilment, calculated cheerfulness or confidence and so forth.

If hope is to be compelling, it must be something more than naïve optimism that things will work out in the manner we desire, because the most superficial reflection reveals that such a disposition is unwarranted. Such 'hope' is often misplaced or frustrated: people don't always recover from illness, injustices multiply, good people suffer and evil goes unpunished. As Roger Scruton notes, 'the belief that human beings can either foresee the future or control it to their own advantage ought not to have survived an attentive reading of the *Iliad*, still less the Old Testament' (Scruton 2013, 2). Nevertheless, Scruton avers, in the very course of recommending a good dose of pessimism, 'I don't go along with Schopenhauer's comprehensive gloom, or with the philosophy of renunciation that he derived from it. I have no doubt that St. Paul was right to recommend faith, hope, and love (*agape*) as the virtues that order life to the greater good' (Scruton 2013, 1). I concur. Hope is, far from delusion or evasion, the virtuous, rational and pragmatic response to a reality that seems to counsel, even demand, despair.

### Desire and Fear, Hope and Despair

To understand why hope is the proper response to our situation, we must begin by distinguishing hope, as I will be using the term, from various forms of desire, with which it is often confused, and by contrasting it with its antithesis, despair.[9] Hope is distinct from desiring, wishing or expecting. While a person might say, for example, that he or she 'hopes' there will be no thunderstorm or that his or her train will depart on time, these are in fact merely analogical and debased misuses of 'hope', just as when a person says that he or she 'loves' ice cream that person is not, presumably, using 'love' in the more specific, and proper, sense of the term.

Hope and despair, unlike desire and fear, have a certain existential quality. By this I mean that hope and despair each assert something about existence itself—the meaningfulness or meaninglessness, the goodness or badness (or, perhaps worse, irrelevance and indifference) of being. This quality is clearest in the case of despair. When a person despairs, truly despairs, he or she loses faith in existence as such. Properly speaking, a person does not 'despair' of the loss of a job, a home or even his or her life (though he or she may fear any of these). To despair is to come to the belief that, as W. H. Auden puts it in his *Twelve Songs*, 'nothing now can ever come to any good' (Auden 1976, 141). This can come about as the result of a devastating individual loss—the death of one's child, for example—but such an event is really just a trigger for a more comprehensive desolation. Despair is not the consequence of the loss of any one particular being or object in the world, but rather of the loss of one's orientation in the world, the loss of one's *sense* of the world.[10] Despair leaves a person with nothing in which she can have faith, nothing to orient or give meaning to her existence, or indeed to existence itself; it is a loss of faith in reality as a whole, the assertion that, ultimately, nothing is valuable.

If despair is the denial of worth, hope is an affirmation that is a response to this denial; it is only in the context of the temptation to despair that we can respond with hope. As Marcel notes, such is clearly the case in our world: 'Despair is possible in any form, at any moment and to any degree, and this betrayal may seem to be counseled, if not forced upon us, by the very structure of the world we live in' (Marcel 1995, 26). William James adds that the 'whole army of suicides' stands as a mute testament to the claim that life is not worth living (James 1956, 34). But it is *precisely* the omnipresent possibility of despair that can become, as it 'was for Nietzsche [and others] . . . the springboard to the loftiest affirmation' (Marcel 1995, 29). Where despair denies that anything is ultimately worthy, hope wagers—hermeneutically, not calculatively—that reality is worthwhile, worthy of my complete engagement (if not, as we will see, my complete endorsement).

Hope, then, rests on a faith; a faith that, despite the apparent lack of worth, the untrustworthiness, the underlying tragedy of reality, there are nevertheless compelling reasons to affirm and endorse it: beauty, goodness, wonder, joy, vitality. The form of hope I am describing does not deny suffering, loss and tragedy, which seem, as far as human experience can attest, utterly inescapable. And while some forms of hope may turn on a salvific doctrine of one sort or another, hope does not require that specific type of faith; it is not, or need not be, a calculation based on some transcendent trick of accounting in which goodness will outweigh, pound for pound, evil, justifying the horror and tragedy. Hope is possible even in the face of the finality of death, even in the face, on a cosmic scale, of heat-death of the universe itself under the influence of the second law of thermodynamics.

As hope and desire have different objects, so do they have different responses. Fear and desire are not blanket positions regarding reality as such;

they are expectant states that focus on the anticipation of some determinate object or event. To desire is 'to desire that X' and to fear is 'to fear that X'. I *desire* that I will be offered the job. I *fear* that the biopsy will come back malignant. Because they are focused on determinate outcomes, fear and desire generally operate in the sphere of what Marcel calls the problematic. Desiring X and not yet having it raises the problem of how to go about securing it. Fearing X and trying to avoid it raises the problem of how to go about avoiding it. As a problematic, desire is tied to what Marcel calls *technics* (Marcel 1995, 30). In the case of problems, the identity of the person seeking a solution is not at issue; it makes no difference who is asking the question—or desiring the outcome or fearing the consequence—because all the relevant facts are external to the questioner. The identity of the questioner can be changed without altering the problem itself; the solution to a problem is, inevitably, a technique that can become common property and used by radically different individuals. 'When I am dealing with a problem, I am trying to discover a solution that can become common property, that consequently can, at least in theory, be rediscovered by anybody at all' (Marcel 1959, 213). If a person desires to lose weight, he or she will employ a technique that is more or less indifferent to his or her individuality. Even if some diet regimes work better for some people than others, any diet one adopts is employed by many other people seeking the same solution to the same problem. So desire seeks some relatively specific object or outcome, which is achieved through the application of some specific technique, a technique that is, in principle, of use to anyone else desiring the same object or outcome.

Hope, however, is of a fundamentally different order: not problematic, but mysterious; not tied to technics, but to a specific sort of hermeneutic wager about being and, therefore, to one's way of being-in-the-world. Mysteries, unlike problems, cannot be solved with techniques, because mysteries—of which love and being are perhaps the archetypal examples—are issues in which the being of the 'questioner' is at stake.

> A problem is something which I meet, which I find completely before me, but which I can therefore lay siege to and reduce. But a mystery is something in which I am myself involved, and it can therefore only be thought of as a sphere where the distinction between what is in me and what is before me loses its meaning and initial validity. (Marcel 1949, 117)

If my being—my unique particularity, my *haecceitas*—is at stake in any mystery, it goes without saying that the response to that mystery cannot be something technical, which is, by its very nature, something indifferent to the being of the person using the technique. My being is not at stake in the changing of a flat tire, which is a problem (an inoperable car) that can be solved with a technique (engage the emergency break, jack up the car, remove the flat and so forth) that is basically the same for any person with the same problem. But when asking whether I love someone or whether my life is worth living, I cannot use a generic technique or universal formula to arrive at a satisfactory answer.

In fact, mysteries are not, properly speaking, the sort of things one 'solves' at all, whether or not the proposed solution is technical. Mysteries do not present us with a gap in our knowledge, or an obstacle to our desires, in the same way that a problem does: 'the mysterious is not the unknowable, the unknowable is only the limiting case of the problematic' (Marcel 1949, 118). Problems are solvable, at least theoretically. For example, the feasibility of economically viable, production-scale fusion reactors is a problem, with a technical solution, whether or not a technical solution is ever discovered. In contrast, the question of love or of the goodness of being is not something I successfully solve once and for all, after which I can turn my attention to other problems. No. A mystery like love is something one must live, day-by-day, in a series of creative acts that help to make the reality of love (Marcel 1995, 34ff.).

Thus, the essence of hoping is not 'to hope that X', but rather simply 'to hope'. That is to say, while hope does not affirm or embrace the current state of affairs—confusion, doubt, depression, infirmity or, in terms of the Anthropocene, climate change—as final, it does not imagine or anticipate some specific change in circumstances that would result in salvation.[11] The hopeful person does, after a fashion, hope for deliverance, but he or she is not fixated on the specific form that this deliverance will take. And indeed, at its limit, as an affirmation of the goodness of being as such, hope does not require such deliverance. But in any case, the more hope transcends any anticipation of the specific form of deliverance, the less it is open to the obvious objection that in many, perhaps most, cases, the hoped-for deliverance simply does not take place. If I desire that climate change will be solved by some particular technique (e.g., economically viable fusion reactors) or state of affairs (e.g., a global awakening to the seriousness of the problem, and a parallel appreciation for the benefits of simpler living, local economies, reduced population and the like), it is highly likely that my desire will be frustrated. However, if I simply maintain myself in hope, without fixation on any preconceived solution or object, then no specific event or absence of an event will have the power to destroy that hope.

Given this account, it would be reasonable to conclude that, in fact, hope is *not* the proper response to the Anthropocene. Climate change is, on most accounts, the final and most terrifying horseman of the Anthropocene. However, insofar as climate change is a determinate *problem*, and is susceptible to technical solutions—whether those technical solutions involve technology (e.g., clean energy, carbon capture and storage technology, etc.) or techniques (e.g., simplicity, vegetarianism, local economies, etc.)—it is the focus of a *desire* or set of desires rather than hope per se. Fair enough. However, precisely because climate change accelerates, magnifies and extends the Anthropocene, it also calls into question accounts of the order and meaningfulness of the world. It is, as I noted above, both a physical threat and an existential/spiritual threat: the age of the Anthropocene is heralded by alarm at the 'end of nature', the 'disenchantment of the world' (Weber 1993) or our loss of our sense of creatureliness (Treanor, Benson & Wirzba 2015). And these narratives court despair as surely as the 'death of God'. Therefore, the novel and terrifying problem of climate change leads us inexorably to the mystery of the Anthropocene.

## Hope in the Anthropocene

Hope, we've seen, is expressed in an affirmation about value of being, an affirmation that bears on my own being and my relationship to the larger whole. As Marcel writes: 'Hope consists in asserting that there is at the heart of being, beyond all data, beyond all inventories and all calculations, a mysterious principle which is in connivance with me' (Marcel 1995, 28). Parsing this sentence will help us to circumscribe the nature of authentic hope. Let's take note of three things: hope involves an 'assertion' that is not based on 'calculation', but instead on a belief about the 'mysterious' relationship between reality and myself.

First, that hope consists in an assertion indicates that it is both a way of viewing things and a way of acting. To hope is not merely to believe certain things about being but to assert, to make active claims about being based on that belief. Therefore, to hope is to adopt a certain way of being-in-the-world; it is an active, not passive, disposition. Marcel makes this clear when he notes that the affirmation of being is 'an affirmation which I *am* rather than an affirmation which I *utter*' (Marcel 1995, 18). Such activity is, no doubt, based on belief, but hope is more than mere belief, which is why I took care to indicate the relationship between hope and faith. Faith is distinguished by, among other things, my willingness to act on the basis of my faith (Treanor 2010).

> Philosophy has, to a large degree, ignored the problem of faith. Faith means belief in something concerning which doubt is still theoretically *possible*; and as *the test of belief is willingness to act*, one may say that faith is the readiness to act in a cause the prosperous issue of which is not known to us in advance. (James 1956, 90)

This criterion seems to apply well in the case hope in the face of environmental crises like climate change. We do not know whether, and we do not know how, we will weather climate change. Therefore, hope cannot have anything to do with certainty; it can be, and often is, plagued by doubt, uncertainty, apprehension, reservations and the temptation to despair. Nevertheless, in the face of this uncertainty, the hopeful person exhibits a willingness to *act as if his or her hope is well-founded*, as if we will, somehow, in some way which is completely unforeseeable (Caputo 1997, 2006, 2013), be delivered from absolute ruin and despair.

Second, the faith or belief that is at the root of hope is not a matter of calculation. With respect to the threats of the Anthropocene, the chances for a favourable outcome are—at least according to many oddsmakers—not good at all. This is perhaps one reason that so many people focus on the *desire* for short-term profit, security or comfort: to ward off the looming *despair* regarding where those short-term concerns are leading us in the long term. When we widen our field of concern beyond the specific problems of the Anthropocene, the impossibility of calculation becomes even more obvious. *Pace* utilitarians, when it comes to the questions of whether being is good, existence is meaningful or life is worth living, we find: (1) that we are incapable of collecting, much less taking into account in a single calculation, the infinitely variegated data points that might bear on the question; (2) that, even if we could enumerate all such relevant data, we would find—precisely because the infinite variety—that we are confronted with incommensurable phenomena that cannot be measured in terms of some universal unit for purposes of comparison and calculation; and (3) that much of what matters when it comes to questions like the goodness of being or the value of life is entirely beyond the reach of facts and data, which certainly bear on, but do not exhaust, the relevant factors in answering such questions.

Finally, Marcel suggests that hope includes a feeling that reality is, in some way, in 'connivance' (*connivance*) with me (Marcel 1949, 68). This is an unfortunate choice of words, insofar as connivance carries the implication of collusion in wrongdoing. True, connivance also carries the suggestion of intimacy, as the etymology suggests a conspiratorial wink as one looks the other way, but hope embodies the togetherness of the 'con-', while eschewing the implicit ignobility of the '-nivance', the *nictare* (wink). Let us therefore amend Marcel's statement to assert that hope finds in realty a mystery by which I am called, among other things, to *collaborate*.[12] That hope demands my collaboration suggests that however hope works itself out, it will require my active participation. That is to say, 'acting as if hope in the future is justified' cannot mean acting as if everything *will* turn out well and so going about my business as if things will turn out well, like a spoiled student failing his university studies because he knows his father will make him vice president of operations in the family business when he graduates. No. Such passive anticipation is expectation or wishfulness, not hope. The hopeful person does not patiently await salvation from some external source, but becomes actively involved in his or her salvation and, in the case of the

Anthropocene, the salvation of the world. So, with respect to hope, a belief gives rise to an assertion, and that assertion is a testimony that I am willing to act on the basis of this belief.[13]

But can we justify such hope?

## THE 'COMPENSATION' OF HOPE

I've suggested that the *desire* for solutions to the problematic aspects of the Anthropocene—for example, a technical solution to climate change—should rest upon a *hope* grounded in the mystery of being: that being is meaningful and good, and that I have a role to play in that story.

One might object that a person could desire a solution for anthropogenic climate change out of pure self-interest without hope in the goodness of being. For some potential climate activists this is no doubt true; the fear of suffering or death in a meaningless and indifferent cosmos might induce one to seek a solution to the problems of the Anthropocene. But that motivation seems unlikely, at least for most of those reading these words. First, as I've argued, the odds are that our efforts will not avoid significant climate chaos and related effects, so people are unlikely to be motivated by the belief that we can actually avoid climate change. Second, those reading this chapter are likely drawn from the lucky minority who will be more resilient to the initial effects of climate change. So a person might think—indeed, many do think—either that climate change is at this point unavoidable, and so without a solution that could be the object of our desire, or that climate change is a problem that will not directly impact their circumstances harshly enough within their lifespan. And if neither a calculated belief in likelihood of successfully avoiding the problems of the Anthropocene nor a personal fear of imminent suffering is an adequate foundation to motivate action, we must look elsewhere for a grounding principle.

It is hope, grounded in a faith in the goodness of being, that can ultimately motivate and sustain the desire to address the specific problems of the Anthropocene. Mitigating and adapting to climate change must *mean something* for us to invest ourselves, and in a world without hope nothing means anything—things can be causes, or effects, or dangers or benefits, but not *meaning*-laden.

Hope is ultimately grounded on what Marcel calls our 'ontological exigence', an aspiration to participate in being and a recognition that being is 'necessary', that is, meaningful:

> Being is—or should be—necessary. It is impossible that everything should be reduced to a play of successive appearances which are inconsistent with each other . . . or, in the words of Shakespeare, to 'a tale told by an idiot [full of sound and fury, *signifying nothing*]'. I aspire to participate in this being, in this

> reality—and perhaps this aspiration is already a degree of participation, however rudimentary. (Marcel 1995, 14)

Ontological exigence is the need, the demand, for some sort of meaning and coherence in the cosmos; it is something like the combination of wonder (*thaumazein*) with an attendant aspiration to understand, if not the entire cosmos, then at least something of one's own place in it.[14] For Marcel, this need is not a 'wish' for being or for coherence, which would be nothing more than a 'psychological state, mood, or attitude a person *has*; it is rather a movement of the human spirit that is inseparable from being human' (Keen 1984, 105).

Erazim Kohák argues that, in our lived experience, the meaningfulness and goodness of being is 'so utterly basic . . . [that it is] never absent from all the many configurations of life's rhythm' (Kohák 1984, 182), even the experiences of suffering and pain (Kohák 1984, 40ff.). As poet Gary Snyder puts it,

> I have a friend who feels sometimes that the world is hostile to human life—he says it chills us and kills us. But how could we be were it not for this planet that provided our very shape? Two conditions—gravity and a livable temperature range between freezing and boiling—have given us fluids and flesh. The trees we climb and the ground we walk on have given us five fingers and toes. The 'place' (from the root *plat*, broad, spreading, flat) gave us far-seeing eyes, the streams and breezes gave us versatile tongues and whorly ears. The land gave us a stride, and the lake a dive. The amazement gave us our kind of mind. We should be thankful for that, and take nature's stricter lessons with some grace. (G. Snyder 1990, 31)

The 'problem of evil', which so many people see as a reason to deny the goodness of being, is, at root, really the 'mystery of the good'. The problem of evil is the problem of why there are pockets or instances of evil in a world that is so obviously, undeniably, primordially good, a world in which we fit, in which we have a place, even if we cannot fully understand it. We have a *natural* affinity for the meaningfulness and goodness of being, which is manifest is experiences such as wonder (Marcel 1995, 13); while such an affinity can be smothered, neglected or denied, it can never truly be silenced (Marcel 1995, 15).

But if goodness 'compensates' us for evil, it is not in some quasi-economic sense (*compensare*, from the Latin to 'weigh together'), in which the scales, on a long enough time horizon, come down in favour of the good—as in Martin Luther King Jr.'s claim that 'the aim of the moral universe is long, but it bends toward justice'.[15] Nor is goodness a justification (*justificare*, 'to do justice to') or vindication (*vindicare*, 'to avenge') that 'conquers' evil by defeating it and doing away with it. Hope does not require theodicy. We don't need to wait for God to put his thumb on the scales of reckoning. In fact, the goodness of the universe and the horror of the universe cannot be weighed on the same scale, as if they were commensurable and able to be measured by some third, neutral unit.

The compensation of the good is, rather, a 'weighing together' in terms of considering together, recognizing the wonder of existence as well as the wreckage and carnage. Poet Jack Gilbert captures this marvellously in his poem 'A Brief for the Defense' (Gilbert 2009, 3). In it, he begins by cataloguing 'sorrow everywhere, slaughter everywhere', 'the suffering [we] have known, and the awfulness of [our] future'. But also he suggests that, despite finitude, despite suffering, despite loss, wonder and beauty remain and, therefore, joy is possible. More, it is an imperative. To focus incessantly on sorrow and suffering is to miss something essential about this life and this world. True, we cannot ignore the sorrow and suffering, which climate change will magnify horribly, but neither can we ignore the beauty and wonder that shine out even in lives extreme hardship (and, given our finitude, all lives eventually come up against extreme hardship). In a manner reminiscent of Virginia Woolf's *To the Lighthouse*, Gilbert recognizes the significance of moments of stillness, beauty and rightness that pull us, momentarily, out of the flow of time in which the second law of thermodynamics has the last word. Such moments, Gilbert suggests, have some compensatory power: 'To hear the faint sound of oars in the silence as a rowboat / comes slowly out and then goes back is truly worth / all the years of sorrow that are to come' (Gilbert 2009, 3).

The world is not 'worth' it because sound of the oars in the silence outnumber or outweigh, genocide, sectarian religious violence, systematic rape, economic injustice and the rest of the litany of human woe, to say nothing of what we've done to the more-than-human world. Nor is it to say that these moments will shield us from further self-inflicted suffering as a result of climate change. The world is 'worth' it because, despite these things, goodness remains and 'hope springs eternal' for those who are in touch with what Marcel calls our ontological exigence.

The goodness of being and the meaningfulness of existence do not rest, fundamentally, on the victory of the good over the horror or indifference of the cosmos in the order of time; hope is, rather, about the victory of the good—beauty, love, life, goodness—in the what Kohák calls the 'order of eternity', that is in the order of value, of meaning, which can become manifest in any moment and to which humans are particular, perhaps unique, witness (Kohák 1984, 82–85, 95–103). Hope does not deny the absolute *reality* of death and dissolution; it denies their absolute *significance*.

When we lose touch with this primordial exigence, we are in danger of slipping into despair. For if being itself is meaningless, despair becomes a powerful, perhaps insurmountable, lure. Those who do not fall into outright despair often resist by shielding themselves, poorly, with a life of mere functionality and *technics* (Marcel 1995, 9–13). Marcel's compatriot Antoine de St. Exupéry captures this well in an account of a pre-dawn ride in an omnibus full of French bureaucrats as they headed toward the then quite dangerous flight to Africa:

> Old bureaucrat, my comrade. . . . You, like a termite, built your peace by blocking up with cement every chink and cranny through which the light might pierce. You rolled yourself up into a ball in your genteel security, in routine, in the stifling conventions of provincial life, raising a modest rampart against the winds and the tides and the stars. You have chosen not to be perturbed by great problems, having trouble enough to forget your fate as a man. You are not a dweller on an errant planet and do not ask yourself questions to which there are no answers. (St. Exupéry 1967, 11)

When it comes to the meaningfulness of being we either (1) hide from the issue by refusing to ask the big questions, burying ourselves in routine, in work, in 'getting ahead' (or just getting by), or we (2) ask the big questions, confront the mystery and conclude either (a) being is meaningful or (b) it is not. If being and existence are not meaningful, then there is not much to do about it other than plug away at whatever will pass the time. But if being is meaningful, then we have a part to play.

## HOPE AS SECOND NAÏVETÉ

As I've noted, I think there are both good reasons to fear the Anthropocene (due to the looming threats of resource wars, climate refugees, agricultural stresses, increased ranges for many diseases, coral bleaching, ocean acidification and so forth) and good reasons to despair in the Anthropocene (because of what it says about our ability to live well as the kinds of creatures we are); I am, perhaps, more inclined than others, to see, and to dwell on, those reasons. Nevertheless, I also think that there is reason to hope, and that based on that hope there is reason to attempt to mitigate the negative impacts of the Anthropocene and to adapt to the changes we've already wrought.

Choosing hope in these circumstances is reminiscent of Paul Ricoeur's account of 'second naïveté'. When a deeply held belief is, as most beliefs ultimately are, subjected to doubt, there are basically three possibilities: scepticism, new belief or a return to the old belief. In the first case, the person plunges into doubt and stays there, remaining, as much as possible, sceptical and uncommitted. In the second case, the person abandons his or her old belief in favour of a new belief, as when a theist, subjected to some dark night of the soul, becomes an atheist.[16] In the last case, however, the person returns to his or her original faith. But having passed through the crucible of doubt, it is impossible to return to the same belief in the same way; this 'return' is what Ricoeur calls *second* naïveté. For example, a child with a childlike belief in an anthropomorphic God who scrupulously rewards good behaviour and punishes bad behaviour might well lose that faith in the face of inexplicable tragedy—say, the death of a parent—only to regain at a later point a transformed, adult faith. Likewise, the dewy-eyed romance of one's youth might be challenged by any number of misfortunes—disease, infidelity or merely the banal routine of paying the mortgage and washing the clothes—only to emerge as a different, more mature love of the same person.

Hope in the Anthropocene seems to me to have a similar form. The innocent faith in progress—the belief that each year will be better than the last and that each generation will enjoy successively greater benefits as we march inexorably, if asymptotically, toward some eschatological promised land—seems impossible to maintain for anyone who is paying attention. Some things certainly have improved in some places, but such progress is neither even nor without other costs, which are all too often misunderstood or ignored. But once a person comes to doubt the future and the direction in which we seem hell-bent on heading—the cornucopian narrative of limitless growth and progress, the goal of independence from the natural world (as if that were possible), the desire to emancipate ourselves from the rhythms and cycles of nature, from our carnal bodies, from finitude and dependence themselves—there is no going back to the first naïveté of utopian belief in inexorable progress and, for many people, despair looms as the likely response, whether it manifests itself at the bottom of a bottle or buried under shiny new consumables and credit card debt.

But there is another option: a return to hope—not the innocent hope of first naïveté, but the transformed, chastened, more prudent, more mature and perhaps somewhat tragic hope of a second naïveté. Such a hope is rooted not in our omnipotent ability to shape reality in order to match and satisfy our desires, nor in the thaumaturgy of salvation by some omnipotent deus ex machina, but rather in a sense of the deep and abiding goodness of being, and a faith in meaning that we can appreciate but never fully understand.

This second hope is neither Pollyannaish about environmental realities, nor a laconic resignation to our fate; it is not a passive expectation for deliverance, whether technological or theological. It requires our active participation in both the enjoyment of the world and the maintenance and promotion of its goodness. E. B. White once quipped, 'I arise in the morning torn between a desire to improve (or save) the world and a desire to enjoy (or savor) the world',[17] which perfectly captures the experience of hope in the age of the Anthropocene. Unfortunately, White also noted that these competing desires make it 'hard to plan the day'.[18]

## NOTES

1. This chapter is an edited version of a piece that first appeared in *Analecta Hermeneutica*, vol. 6 (2015).

2. Or 390 BC, depending on the source.

3. The Anthropocene is a term that has gained popularity over the latter half of the twentieth century. It designates the geological timeframe marked by the domination of human influence on the global environment. If the Anthropocene is characterized by the dominance of human influences on the planet and its systems, then surely climate change is the exemplary case, insofar as it has changed the terrestrial climate, as well as the temperature and pH of the ocean, there is now no corner of the Earth untouched by humans. Bill McKibben, one of the earlier popular writers to draw attention to climate change, goes so far as to suggest that it marks the 'end of nature' (McKibben, 1989).

4. 'Benign demographic transition' is a phrase capturing the tendency for increasing affluence to lead to more environmentally beneficial behaviour—lower birth rates, increased environmental awareness and concern and so forth. My phrase, '*malign* demographic transition', captures the other side of the coin of development. As people become more affluent they consume more, and because consumption, especially modern consumption, is steeped in fossil fuel at every stage of production and transportation, the environmental effects of development have a very obvious dark side.

5. Although we have caused climate change, it came about as an unintended and largely unrecognized consequence of otherwise innocuous actions. The development of the coal-fired steam engine in 1712 allowed us, for the first time in history, to draw on seemingly unlimited solar power that had reached the Earth millions of years ago, now stored in the form of fossil fuels, rather than much more limited solar power that reached the Earth in the last year or so, growing the plants that fuel both human and animal muscle power. Harnessing that stored power was a revolution—one with unintended consequences. It was the massive release of this carbon into the atmosphere, largely unrecognized at the time, that began anthropogenic climate change. Because our entire way of life has adapted to cheap, dense and easily portable forms of fossil fuel energy, basically everything we do contributes to climate change.

6. The nature/culture distinction is, of course, hotly debated among philosophers and other scholars. However, covering this important debate would distract us from the topic at hand.

7. I often tell my students that (1) a responsible future would include pulling back from consumption of all sorts, reducing travel and the use of fossil fuels, localizing economies and so forth, but (2) given our inaction, indolence and selfishness, it is unlikely we will manage to 'power down'. Thus, there are (3) two possible futures for humanity: one looks like Cormac McCarthy's *The Road* and the other looks like *Star Trek*. But there is no real justification, other than a leap of faith, for thinking we will be living in a *Star Trek* world before climate change surges past any number of tipping points, which means there is some likelihood that we will be living in a radically impoverished environment (at least as viewed from the human perspective).

8. As well as the intransigence of the laws of physics and chemistry, which do not compromise or negotiate.

9. My account of hope here is, quite clearly, deeply indebted to the work of Gabriel Marcel. However, my own account differs from Marcel's in important respects, and so this should not be taken to be a strictly Marcelian description of hope.

10. 'Sense' here taken in the way it is used by Erazim Kohák, as a kind of foundational, global impression based on experience (and, later, reflection). 'Philosophy can claim to be the *scientia generalis* because it seeks to see and articulate the sense of being as it presents itself primordially, prior to the imposition of any special perspective or purpose' (Kohák 1984, 49).

11. Salvation, for the time being, taken in its broadest sense as deliverance from an unwanted situation or state of affairs.

12. More on this below, but note that the gesture toward a 'mysterious principle in connivance with me' does not demand belief in a personal god; it can just as well be the recognition of meaningful being of which I am a part. Although both Marcel and Kohák argue for a

personalist account of being, the latter especially is quite careful to articulate his account in a way that is not tied to any specific, determinate faith tradition.

13. Of course, neither one's faith in reality, nor one's action on behalf of that faith can assure a positive outcome. Just as hope is only possible in the face of the possibility of, and temptation to, despair, so faith is only possible in situations where it (the leap of faith) can prove unfounded—one cannot have 'faith' in a certainty.

14. Marcel concedes that it may be the case that ontological exigence is never fully satisfied, and cannot be fully satisfied (Marcel 1973, 50).

15. Martin Luther King Jr., *Sermon at Temple Israel of Hollywood*, delivered 26 February 1965.

16. Although here, as in all cases of second naïveté, effects of the first belief persist. Ruptures with our past can be more or less radical, but they are never absolute.

17. 'E. B. White: Notes and Comments by Author'. Interview with Israel Shenker in *The New York Times*, 11 July 1969.

18. Ibid.

*Chapter Eight*

# Have Hope, Not Too Much, Mostly for Plants

## *Hope in Environmental Moral Literacy*

Jonathan Beever

I argue, in what follows,[1] that hope is a danger to ethical decision-making in environmental ethics and policy, and that this danger has been largely overlooked within the complex conceptual apparatus of ethical decision-making. First, I articulate the conditions of environmental duress under which younger generations have grown up and the potential that hope has to shape future decision-making. I then analyse hope's place in the context of moral literacy, or the learning framework for ethical decision-making. From that framework of moral literacy, I then outline conceptual problems with hope, emphasizing particularly hope's complexity in concept and application. I propose that the claim that hope (and other character virtues) is necessary to the ethical decision-making process is made uncritically. Instead, the significant epistemic complexity and thick concepts of environmental ethics alongside the conceptual complexity of hope itself deny hope a clear target, and so deny its necessary role in ethical decision-making. Intersection with the literature of bioethics is informative here. Yet hope continues to play a practical and initial role in motivating action. This conflict, between the practical role of hope in ethical decision-making regarding the environment on the one hand and the fuzziness of the target of hope on the other, leave hope in the precarious but important position as inspiration to consider problems of environmental value in the first place. In the last section, I argue that this conflict opens space for the manipulation of hopes related to environmental ends. I conclude that if we take hope to be necessary for ethical decision-making, then we ethicists ought to tread very carefully around its analysis and future application.

## UNDER CONSTANT DURESS

The natural environment is in trouble, we are constantly reminded. For people of my generation (pre-millennials) and subsequent ones, this has been the case *for our entire lives*. We're too late to prevent significant climate change, having surpassed the four hundred carbon dioxide parts per million marker as of 6 May 2015,[2] too slow to have prevented the eradication of nearly half of all wildlife species in just the last forty years and too apathetic to change a political landscape that seems not only unconcerned but downright vicious in its continued neglect of environmental systems. So bad are conditions that cultural critics are marking a shift in young citizens, like the 'new fatalism' (Hoffman 2015) of young Japanese. In early 2015, the weekly magazine *Shukan Kinyobi* reported evidence of an interesting paradox: young Japanese despair the state of the world and yet report being happier than ever (Hoffman 2015). This rather strange result is explainable, to some, by understanding 'happiness' in darker terms. As journalist Michael Hoffman reports, darker happiness may be a reinvigorated fatalism: 'the contentment the opinion polls consistently find in the young generation is mere resignation' (Hoffman 2015). Anecdotal evidence from U.S. environmental educators I correspond with suggests a similar trend in younger generations of U.S. students. As Lisa Kretz reports, marine ecologist and TedX'er Elin Kelsey has found further evidence for this trend. 'Elin Kelsey met with children aged [ten to fourteen] years old from [ninety-two] countries (ranging from the very impoverished to the very wealthy) at a United Nations conference. . . . When questioned regarding their feelings about the environment the children's responses, rather than ones of wonder, were of despair, anger and worry' (Kretz 2013, 938). Whether from resignation or ignorance, trends in young attitudes about the environment like these suggest either fatalism or apathy, and they deny the sorts of climate education and activism that can cultivate change in climate—environment and cultural.

Indeed, since at least the birth of the Generation X-ers, children have grown up knowing nothing except that the environment is in trouble. Thanks to Rachel Carson's research, the resurgence of fascination with Aldo Leopold's writings and the popular rise of the American environmental movements in the late 1960s more generally, entire generations of Americans have grown up in a world where the environment *has always* been in trouble. Members of these generations have known no other social and environmental reality, if they know any. Species extinction and biodiversity loss, human-caused changes to climates and new forms of pollutions mark the sort of world in which they live. Those of us particularly attuned to environmental thought, like the readers of this book, likely recognize this social and ethical framework and can analyse its historical and genealogical place within a larger context. But, I suppose, we are in a minority in this ability. The majority of individuals exist in *umwelten* uniquely shaped by this hermeneutic, unable to see or at least untrained in seeing the world in other terms: they see a world under constant duress and are developing a deep-seated fatalism in response to it.

With things apparently this grim, it would be a simple thing to throw up our hands in anger, despair and gloom and to abandon hope for a better future. It is no surprise then that the contemporary feelings about the environment of children and youth are similarly ones of despair and anger, rather than wonder or joy. Kelsey identifies what she calls 'a real narrative of gloom and doom' (Kelsey 2012) but offers an optimistic message. Blowing bubbles to represent the bubble nets blown by collections of humpback whales to drive prey together, Kelsey calls her audience to build up a collective power to positive action (Kelsey 2012) by working together. Now, of course, we might fault optimistic bubble-blowing educators like Kelsey for neglecting the seriousness of the environmental crisis, or crises, and, likewise, we might praise her for that same optimism because it offers a message of encouragement—and of *hope*. Indeed, Kelsey's ultimate call is for researchers and environmentalists alike not to cultivate a cautious optimism but instead to spur on a full-on 'torrent of hope' (Kelsey 2012). Philosopher Lisa Kretz finds kinship with Kelsey's message, articulating a view that hope plays an important role in that it gets 'us motivated to act in ways that help bring about the possibility of better states of affairs' (Kretz 2013, 942). Our task, as environmentally oriented educators and policy leaders, is to cultivate an environmental and ecological hope that is 'infectious and can be fostered through teaching' (Kretz 2013, 942). A torrent of hope, in this view, leads to a self-perpetuating cycle of justifying performed actions in support of environmental goals by more and more individual agents.

On my reading, however, spewing forth a torrent of hope is a dangerous move. As I've argued elsewhere, incautious hopeful optimism can make one act the ass:

> Given that bleak landscape of environmental duress, it is easy to take the view that for each small happiness and every personal triumph there is a boundless host of uncertainties, doubts, fears, pains, and sorrows waiting to follow close behind. It certainly seems that some take comfort in donning blinders that draw attention only to the occasionally offered oat bucket of happiness while masking the inevitability of the slaughterhouse in which they are served. The new job (hooray!), the published article (congratulations!), the birth of the new baby (way to go!); the rekindled friendship, the fiery romance, the thoughtful note, the kind word or loving touch, the budding flower, the sunrise over a new

> day, the sound of a cool breeze coming after a hard heat: these are the oats that sustain the hopeful ass. But like happiness, oats, digested and effused or unnoticed in the corners of the bucket either turn sour or dry up, depending on their environment. (Beever 2014)

My allegorical soured or dried up oats represent the bland optimism that I see as the danger of unwarranted or uncritical hope. Yet, as psychologist Per Espen Stoknes argues, it is important to work to separate hope from bland optimism (Stoknes 2015a). Justifying and performing actions and attitudes that are ethically appropriate requires much more than hope: it requires what I take to be an impressively robust process of ethical decision-making of which hope is just one small component. A torrent of hope threatens to drown out ethical decision-making in a puddle of the blind incautious optimism of hope's simulacrum.

## HOPE'S ROLE IN MORAL LITERACY

> 'Hope, then, is the principal antidote which keeps our hearts from bursting under the pressure of evils.'
> —Ferguson & Allen 1880, 465

So what would it take, then, for ours and future generations to make ethical decisions regarding the natural environment? This is a question not only of education but also a question of inspiration and motivation. Constrained by a view of the natural environment that suggests that things are inevitably bad and getting worse off, an individual's capacity to make ethical decisions (particularly those related to environmental considerations) seems to me necessarily impoverished, compared to those not so constrained. Epistemic commitments as well as ethical values of such individuals are impacted by embedded normative orientations to the world such as that deeply pessimistic or fatalistic view of the future of natural environments.

Yet many environmental educators—myself included—continue to trust that it is possible to shift that embedded orientation, thereby opening space for making ethical decisions in different contexts. Besides cultivation of a deep understanding of ecological relationships, environmental value(s) and frameworks for understanding, such a shift would take inspiration to thoughtful inquiry about issues of ethical importance and also motivation to develop that inspiration into informed ethical action. Each of these components to making ethical decisions is markedly complex, especially in the face of constant reminders of ongoing threats to environmental health, biodiversity and ecosystems resilience. And even if open to considering environmental challenges, ethical decision-makers question what conclusions they should draw from that decision-making process, faced as they are with a world under duress. One key component of answers to such questions has involved the cultivation of certain individual character traits, or virtues, like hope.

Several scholars, like Kelsey and Kretz above, have argued that hope plays a significant role in motivating ethical decisions. Indeed, hope may even be necessary—if not sufficient—for ethical action. However, few scholars have articulated any coherent framework for ethical decision-making, and hope's role in such a framework has been taken as a rather uncritical assumption.

Within the context of ethics education, philosopher Nancy Tuana continues to work on the development of just such a framework. Tuana, like Kelsey and Kretz and others, thinks that Hope is important to ethical decision-making. Within the framework Tuana works to articulate, virtues play a framing role in ethical decision-making. For her, hope—in this case *moral* hope—is required to initiate the asking of ethical questions and the identification of ethical issues.

For Tuana, *moral literacy* reflects the idea that 'the skills and knowledge specific to making ethical choices in life are learned capabilities requiring skills in which individual can be more or less competent' (Tuana 2007, 365). To be morally literate, like being literate in the sense of being sufficiently able to read and write, requires an ongoing education or development whose goal is the cultivation of a particular skill. The development of a skill like either reading literacy or moral literacy takes a considerable amount of time, energy and mental reflection. That skill then becomes slowly habituated over time into consistent action and *reflexive*[3] thought, *ceteris paribus*. Indeed the *other things* here are centrally important: social and environmental factors can have a substantial impact in the extent to which and consistency with which literary habits are upheld and cultivated. One need only consider socioeconomic disparities and reading literacy rates (Van Vechten 2013) to recognize such an impact. Although there is less empirical evidence in support, it is easy to extrapolate similarities here between these disparities and the cultivation and maintenance of moral literacy habits.

In her developing work, Tuana (2014) argues that hope is an important component of moral literacy: part of a package of motivating moral virtues that, on her account, enable the enactment of moral thought in action. Moral literacy, in her view, consists of three major components: ethical sensitivity, ethical decision-making and ethical motivation. Ethical sensitivity, drawn from the literature of Kohlbergian moral psychology (see Bebeau, Rest & Yamoor 1985), is the development of perspective-taking skills that enable the morally literate decision-maker to interpret 'who is involved, what actions to take, and what possible reactions and outcomes might ensure' (Tuana 2014, 158). Ethical decision-making, the second of these three components of moral literacy, involves a cognitive set of skills enabling the morally literate to understand, apply and evaluate various ethical frameworks, relevant facts and options for action (Tuana 2014, 165). This cognitive process has traditionally been taken as *the* relevant process by which individuals make (hopefully ethical) decisions when confronted by an ethically loaded problem. Yet to assume that this cognitive skill was always or even often completed rationalistically would be to ignore the growing understanding of the important role of affective components of the ethical decision-making process. So, Tuana's account of moral literacy includes its third component: ethical motivation.

Ethical motivation is driven by *virtues*—or 'types of moral habits and sentiments' (Tuana 2014, 169) like courage, hope and purpose—that combine to drive thought into action. Tuana notes that, 'Indeed, moral courage goes hand-in-hand with moral hope and motivates us to risk hardships to do what is right and attain what is just. Without hope, we experience despair or cynicism, and perhaps even indifference' (Tuana 2014, 172). Hope, on this account, plays a central role in the story of ethical motivation by enabling or motivating ethical action. Hope, along with courage, is the sort of human characteristic that catalyses the processes of ethical decision-making and action-taking. These characteristics drive, affectively, the otherwise rationalistic process of evaluation and assessment of options. Each of us, as moral agents, *thinks*, *feels* and *acts* in cultivating our individual moral literacy. This account is not particularly novel, since the interplay between cognitive and affective states is a central focus of twenty-first-century philosophy, from neuroscience to moral psychology and epistemology. My purpose in developing it here, however, is to reiterate just how central it is to contemporary ethical problems—especially those regarding the natural environment—and also to assert its complexity. Hope, as a factor in moral literacy, is compound in its complexity: as a concept, in relation to other concepts, and in its application to environmental problems.

## THE PROBLEM(S) OF HOPE

> 'Used with a due prudence hope acts as a healthful tonic; intemperately indulged, as an enervating opiate.'
> —Ferguson & Allen 1880, 466

On accounts like the ones I've explored above, moral hope plays a vital role, by fiat, in ethical decision-making. Specifically, it *enables* or *motivates* the decision outcome part of ethical decision-making. Without it, ethical decision-making would remain a hypothetical process. Imagine, for example, a couch-potato sort who thinks to himself: 'There is surely something better on another channel. But the remote is out of reach, so I suppose I'll just watch this terrible television program instead'. Analogously, our couch-potato moralist might well understand which decision is the ethical decision in a particular situation and yet fail to act: 'I suppose the right thing to do is X. If only I had the motivation to do X. Instead, Y seems easy, so I'll do Y'. So, hope plays an important motivational role. But this important role is confounded by complexity of the concept and many of the ethical problems to which it is tasked with motivating responses.

Hope's complexity is evidenced first in the expansiveness of its use as a term and concept. Recognizing this expansiveness is the first step in determining what type of hope is relevant to environmental ethical issues. While I will focus on hope's complexity as a problem for its use in ethical decision-making, I think hope faces other important problems as well, including the nature of virtues (what they are, ontologically: states? traits? behaviours? etc.) and their compatibility (or lack of compatibility) with other ethical theories and principles. Even if, in the end, we agree that hope's complexity is a resolvable problem for ethical decision-making, these other conceptual problems remain unaddressed.

On Tuana's account of moral literacy, moral hope—or hope that is aimed at the achievement of moral purpose (Tuana 2014, 172)—is of a different kind than other sorts of hope, like hope for a bright future or hope for recognition of good work at one's job (Tuana 2014, 172). These other sorts of hope, while evidently still habits or sentiments, are directed toward non-moral ends. Getting job recognition or having a bright future are instrumental ends—the sorts of thing we might want for ourselves but not the sorts of things that are moral ends in themselves. There's a complicated metanormative account to be offered here: why think, for instance, that having a bright future isn't just the sort of thing that would be morally good in itself? If it is that sort of thing, then why shouldn't we consider it a moral hope? Part of the answer is that moral hope, on Tuana's account, is importantly linked to action. 'Hope', Tuana describes, 'is not an abstract concept or merely a belief, but a state of being in which we are called to action. Moral hope is not a belief, but a disposition to act' (Tuana 2014, 172). So I could have any number of hopes. I could hope for a bright future full of wondrous things or I could hope that my financial investment doubles in value in fewer than five years. Those hopes might even have moral content: they might have to do with a normatively better state of affairs instrumentally or intrinsically. So moral hope has a two-part definition involving, first, moral content, and second, motivational capacity. Thus unless my hope somehow drives action—and, further, action toward some moral end—then it is not moral hope.

So, ethical decision-making is governed by a particular type of hope; namely, moral hope. A second sort of complexity to address is the *value* of moral hope: if it is indeed a characteristic we ought to cultivate for its essential role in ethical decision-making, then it must be valuable. This sort of question about the nature of moral hope follows Luc Bovens' (1999) appeal to a two-level story of the value inherent in hope. First, Bovens argues that hope's value lies, in part, in the role it plays in offering respite in times of hardship (1999, 676). The view here is that the gloomy or fatalistic youth can still find a spark of environmental optimism even in spite of dramatic environmental damage by holding onto hope. It seems to me that, within environmental philosophy, this is a rather rose-coloured view. Holding hope up as a respite in times of hardship sounds suspiciously like providing false relief in denial of inevitable harms and consequences of previous actions—like the scaffolding risk of releasing carbon dioxide into the atmosphere. Maybe a little childhood gloom is an appropriate response to being handed a world that has been deeply harmed by previous generations.

The advocate of hope might, perhaps, fall back on an appropriateness claim and defend the role of hope by suggesting that an appropriate level of respite will prevent psychological miring, or stuckness in inaction that might come along with an overwhelmingly negative perspective on the state of things. Bovens' second claim is that hope also has value insofar as it leads to the restructuring of second-order more general hopes. 'It is also conducive to an increased self-understanding', Bovens writes. 'As I shift my hopes to more attainable and meaningful pursuits that are no less constitutive of a better professional life, I have come to learn something about myself and my place in the world' (Bovens 1999, 676). This second claim is more centrally relevant to environmental thought: even rose-coloured hope can help us reconstitute new hopes more relevant to our place in the world. I may initially hope that the natural environment I grew up in remains exactly as it was when I was young. However, as I learn and my epistemic and ethical value perspective shifts, I may come to hope, instead, that ecosystems to which I'm related remain resilient in the face of inevitable change. Here, perhaps, is the response to my claim of hope's providing false relief. The work of the concept, in ethical decision-making, is to be done at the level and extent of application: too much hope is naïve and potentially harmful, and too little hope breeds inaction and morosity. Hope's value, in Bovens' view, is at least two-fold: as a respite during 'dark times' (Jamieson 2014) of environmental duress and as a source of second-order sustainability of other related hopes. But, the value of hope is itself complex: its value is contingent, it seems, on the sorts of problems to which it may apply. Perhaps contemporary environmental issues are dark enough not to be worthy of hopeful optimism, such that hope has little to no value at all in the arena of ethical decision-making.

There is yet another level at which a hope's complexity can be addressed. In addition to the *type* of hope and the *value* of hope, hope can be differentiated in terms of its *directionality*. On the one hand, we can cultivate a *forward-looking* hope. Hope, oriented toward a better future state of affairs, serves the purposes outlined above, leading the hopeful individual to more attainable and meaningful pursuits and compelling ethical decisions into ethical action. Psychologist Per Espen Stoknes defines forward-looking hope as 'active' or 'heroic' hope. For Stoknes, forward-looking hope is characterized by optimism about future states of affairs: 'This type of optimism says that the likelihood of a good outcome depends on the magnitude and acumen of our effort. It may be a fight, but one we're going to win' (Stoknes 2015a). A forward-looking hope is, in some sense, historical, genealogical and critical. It is critical insofar as it problematizes past determinates of some current state of affairs, either through historical or more richly genealogical methodology. This hope looks forward in a historical and normative sense, motivating action that would encourage or implement a better state of affairs for the current context.

On the other hand, we can cultivate a *beyond-looking* hope. Hope that looks beyond (rather than forward) is blind, ahistorical, linear and noncritical. It orients the hopeful away from the current context and toward some transcendent other-than-this perspective. Stoknes describes beyond-looking hope as 'passive hope' (Stoknes 2015b). Its often religious context offers an orientation toward some normatively better future state of affairs largely out of the control and purview of the individual agent. 'It is an outlook', Stoknes writes, 'where—if you think positively—all is sweetness and light. Because the world turns out well anyway, there is no reason to worry and work; we can wait for rewards to ripple down our way' (Stoknes 2015a). In the post–World War II German tradition, where hope played an important role as thinkers began to think past and beyond nationalism and identity politics, Ernst Bloch wrote against this sort of beyond-looking hope. Bloch, in his *The Principle of Hope*, worked to 'teach humanity to learn hope, as opposed to the vortex of fear that leads into is dominated by nihilism. . . . The positive appraisal and exposition of the "dream", supremely the "daydream", is the first point of entry into a conception of hope uncorrupted by the illicit "loan" of the idea of transcendence as an otherness beyond the human condition' (Roberts 1987, 94). Bloch was critical of this move toward transcendence, toward a looking beyond, in discussions about hope. For Bloch, while hope had a vital role to play in freeing up human action in the face of nihilism, it could not fulfil its role in the form of a looking beyond. In our contemporary landscape, a beyond-looking hope runs the same risk. It runs the risk of divorcing the ethical content and context of decisions from the orientation toward action. If hoping for some transcendence source or relief or some improbable future state of affairs overcomes taking action in response to immanent threats to health and well-being, then it has failed its roles both as Tuana's ethical motivator and as Bovens' moral reconstituter.

Even in this distinction I have made between forward-looking and beyond-looking, there are additional distinctions to be made. For example, Stoknes' own argument is on behalf of a third option, what he calls 'grounded hope' (Stoknes 2015b, 222). Grounded hope is based not on future outcomes (whether they be forward- or beyond-looking) but rather on individual character. To have grounded hope is to be actively optimistic without expectation. Stoknes notes, 'I don't need to believe that things will end well in order to act. The walking and the doing are their own reward' (2015a). The active scepticism (Stoknes 2015a) involved in this view encourages the individual to act not because the individual *knows* that action will influence or predict the future but, rather, because the individual *hopes* that action may so influence. Grounded hope is a hopeful optimism that humanity will survive whatever future states lie ahead by simply continuing to move forward. In an effort to avoid the obvious criticism that grounded hope engenders naïve optimism or, worse yet, the fatalism described earlier in this chapter, Stoknes supports grounded hope with a gesture toward the nature of epistemic uncertainty. 'Nobody really knows enough to be an absolutely convinced pessimist', he writes, 'Sure, things may look bad, and optimism may seem rationally impossible to me today, but that doesn't necessitate flicking the switch over to pessimism' (Stoknes 2015a). The complexity of grounded hope comes precisely from the epistemic complexity of assessing future states of wildly complex interlocking systems like those making up the natural environment. Whether forward-looking, beyond-looking or grounded in present action, hope's directionality is a third marker of its complexity.

The frequency and range of usages of hope differ (at least) in terms of the type (moral or nonmoral), value (intrinsic or contingent/instrumental as motivator and/or encourager) and direction (forward-looking, beyond-looking or grounded). In applied contexts, this complexity often clouds the strength with which analysis of hope's role and impact can be coherently completed. Reliance on a rather vague notion of hope haunts the bioethics literature, for example. Bioethicists Coeckelbergh and Mesman, for example, conclude their 2006 paper by stating that 'it is up to [practitioners], as individuals and as a team, to re-imagine the practice of intensive care, and to continue to make their daily decisions with hope and imagination' (2006, 20). Individual practitioners—let alone individual philosophers and theorists—are tasked with utilizing hope in re-imagining care. Little serious conceptual work is offered in that essay; rather, the authors list hope among a range of other emotions including fear and doubt as factors impacting how imaginative reasoning plays out for patients and practitioners in the context of ethical decision-making.

For those of us interested in hope's role in environmental ethical debates, the broader bioethics context is instructive. 'There is no consensus on the meaning of hope', some have pointed out (Patton et al. 2013, 90). Yet hope continues to play an important role in bioethical discourse between physicians and patients. Physicians are regularly faced with the ethical dilemma of offering medical truths to patients and offering shades of truths: both of which may leave the patient haunted by spectres of doubt and fear about health outcomes. In current medical care and medical ethics, medical objectivity—warranting rather brute honesty in relaying health information to patients—is weighed carefully against constructive patient communication. Patient-centred care emphasizes empathetic listening and discursive practice pushes physicians to moderate the transfer of potentially upsetting or 'bad news' information to patients by offering, in addition, some reason to remain positive, or to *have hope*.

Further, in the biomedical context there are several pragmatic and health-related reasons to give a patient hope. Psychological well-being that comes from hopefulness has been evidenced to increase positive health outcomes, or at least alleviate psychologically negative outcomes of negative health prognoses. Having a self-conception of one's health is linked to physical health implications: psychological and physical health are correlated if not causally connected (e.g., Collingwood 2015).[4] Contemporary literature on hope in a biomedical setting has tended to focus on the need for a careful balance between false or empty hope (as in 'getting one's hopes up') and positive hope (as in 'have hope for the future'), analysed quantitatively.[5] But the focus on hope remains as grossly undertheorized as it has remained in the broader context of ethical decision-making: there is no coherent conception of the nature, value or direction of hope. Nor is there clear conception of the role it plays in medical practice, nor of the role it *ought* to play in medical practice. Yet, its importance remains an assumed premise in the arguments about the nature of ethical decision-making.

A relevantly analogous scene can be set in environmental philosophy as well, where ethical dilemmas, for instance, between reporting of 'raw' climate science data and careful representation of complex results to a nervous public. But in the context of environmental ethics, the epistemic complexity of interconnected natural systems is a further compounding factor. Take, as an example, the recent debates about de-extinction of the woolly mammoth (Brand 2013). For environmental activist Stewart Brand, who had spearheaded much of the support for de-extinction programs, bringing back the woolly mammoth would reinspire 'the awe and wonder that drives all conservation at its deepest level' (Brand 2013) building and building on hope. And yet, de-extinction of the woolly mammoth might well have negative broader ethical impacts; for instance, on the welfare of individual animals it has 'brought back' or by detracting from work on sustainable current biodiversity for future generations. Simply understanding and recreating the genomic structure of the woolly mammoth (Ghosh 2015) will not bring it back. The mammoth was linked to particular ecosystemic conditions, along with historical social systems, and a broad range of other environmental interactions and functions. The epistemic complexity involved in a robust understanding of what it means to 'de-extinct' the woolly mammoth is immensely complex and coupled to equally complex ethical conditions. Yet, for Brand and other de-extinction proponents, the central importance of this work is in building up *hope*: offering impetuses for the feelings of awe and wonder that make us optimistic about the future of the natural environment and its species, past and present.

We can imagine any combination of type, value and direction of hope in cases like this, but here let's focus on two seemingly polar approaches. First, consider moral grounded hope taken as intrinsically valuable. In this sense, moral hope still motivates ethical action but does so through a carefully sceptical analysis of best options. No pre-given external or future goal orients ethical analysis; rather, ethical decisions are made based on current empirical information and ethical constraints—much akin to the process of optimization familiar to engineers. Next, compare this vision to moral hope taken as instrumentally valuable and beyond-directed. Here, the goal of action is predetermined, its value set by some external and abstract condition or valuer. Further, these goals are not *merely* valuable—like the goal of making more money in a future job—but *morally* valuable: they are oriented *rightly* and toward goodness. Finally, the moral worth of hope is instrumental to these broader ethical ends: hope is a tool used to get to broader ethical ends. These two approaches appear, to me, grossly different with the latter of the two being practical and the former an ideal to which the ethical decision-maker (and theorists) strive. Also, importantly, the latter approach to hope—beyond-looking and instrumentally valued—opens dangerous space for the manipulation of environmental ends. In the last section, I examine this dangerous space and argue that hope, especially in the context of environmental ethical decision-making, deserves more careful analysis.

## HOPE AND THE MANIPULATION OF ENVIRONMENTAL ETHICS

> There are so many humiliations in this world! The secret is to rise above them, to throw off dissatisfaction, and to grasp some pleasing hope, grateful and beneficial to the mind. We are encompassed by illusions and delusions. We need the comforting promises of the heart—a steadfast faith in the good and true, and hopefulness in all things, especially of futurity. Hope is rich and glorious, and faithfully should it be cultivated. Let its inspiring influence grow in the heart; it will give strength and courage. . . . The way of the world is dark enough even to the most favored ones among us. Why not, then, gather all the happiness out of life that you can? Why not strive to cultivate the cheerful, hopeful disposition that will enable you to see the silver lining to every cloud?
> —Ferguson & Allen 1880, 471–72

Ferguson and Allen, in their last nineteenth-century home classic *The Golden Gems of Life* (1880), propose that having hope is a tonic to the duress and dissatisfaction of living. Their proposal, as I've suggested above, is a naïve proposal. It ignores the complex role that hope plays in the larger context of moral literacy. It fails to account for the importance of facing some problems—like contemporary environmental problems—with a greater degree of realism. And it denies recognition to the ways in which the hope can be manipulated to redirect or revalue some ends of environmental ethics. Their distinction between hope as a 'healthful tonic' and as an 'enervating opiate' (Ferguson & Allen 1880, 466) parallels the classical Greek understanding of Hope's mischievous nature.

In the classical Greek tradition that is her origin, Hope (*Elpis*) is a sprite: a mystical tease who offers mere glimpses of an idealized future which is constantly and inevitably just beyond reach. There are no logically sound arguments to be made in her favour, no empirical justifications to be offered as to her existence from the 'great blooming, buzzing confusion' (James 1890, 462) that is the natural world. Hope—as a sprite and as a virtue—is a myth in the sense that she is placed before us a priori. The Greek term ελπις only means hope insofar as hope is a specialization of 'expectation' that in turn is a specialization of 'supposition' (Verdenius 1985, 70). 'Hope', then, points to an expectation: whether of a good or an evil (Verdenius 1985, 66). So the term hope, for the Greeks, already indicates the uncertainty about the future that comes with its complexity in concept and in application.

As a concept, hope is richly complex in terms of its type, value and direction. In its application, hope is epistemically complex as we have seen above in the case of the proposed de-extinction of the woolly mammoth. As concept and in application, the epistemic complexity of hope is inextricably coupled to equally complex ethical conditions. Ethical values are deeply embedded in the beliefs, attitudes and actions of every citizen, scientist and policy-maker. Given all this complexity and uncertainty, hope's role in ethical decision-making makes the decision-maker vulnerable to manipulation. If I could convince you, for example, that bringing back the woolly mammoth would inspire you to hope for more and future action in support of ethical environmental ends, I would have fulfilled moral hope's instrumental role. You would, I might hope, be motivated to engage more actively and often in ethical decision-making toward such ends and, in turn, cultivate a stronger sense of moral literacy. But this inspiration is a form of manipulation. I would only have to shift my intentions—and realign my own moral ends—to inspire potentially wildly different ethical motivations in you. Perhaps, for example, I could inspire hope in you about the future potential benefits and safety of deep-water offshore oil wells. Perhaps, with enough hope, you would become motivated to belief and then act toward *those* ends.

Of course, your own motivations are shaped not by my inspiration alone but by a wide range of experiences, both internal cognitive and affective experiences as well as external interventions. The extent to which manipulation of ends is possible is correlated directly to the *vulnerability* of individuals with whom I interact. Vulnerability has been an ongoing topic of interest in the bioethics literature. Many scholars have articulated the role that physicians and medical practitioners play in shaping the imaginations and hopes of patients, who struggle with uncertainty of outcomes and complexity of conditions (e.g., Simpson 2004, 434) and are vulnerable to exploitation (Martin 2008). Bioethical vulnerability accounts for the ethical and epistemic complexity of medical decisions. Environmental vulnerability can be understood analogously as the state of individuals under conditions of environmental duress and facing the ethical and epistemic complexity of globally complex environmental problems. Hope, as a necessary condition for ethical decision-making in the framework of moral literacy, can motivate decisions either ethically, by valuing hope itself as a motivation factor toward grounded ends or unethically, by manipulating values and motivations toward selfish ends. Ongoing work needs to be done to differentiate 'grounded' from 'selfish' ends under the conditions of any given case and to reach consensus about environmental goals more generally.

In this chapter, I have conceded that hope is necessary for ethical decision-making insofar as it motivates or inspires individual consideration of environmental ethical ends in the first place. But hope's complexity as a concept and in application makes the decision-maker vulnerable, distracting us from the actual hard work of decision-making and allowing us to ignore the vastness of epistemic and ethical constraints at work in each decision. Hope can bypass the discomfort of human emotional distress, avoid the hard work of reasoned and reasonable thinking and ignore the tiring process of engagement with complex communities of moral agents and moral patients. Further yet, it is open to manipulation by those who might wish to shift or alter the proposed ends of environmental action. Hope's complexity, coupled with its importance to ethical decision-making, must be recognized and thoroughly analysed. Without this level of responsibility regarding hope, the project of cultivating moral literacy is incomplete.

Returning to the apathetic and fatalistic condition of recent generations with which I began, we see that moral hope is supposed to be a capacity or habit that—when properly developed and understood—can lift these and future generations out of ennui and back into a position of awe and joy regarding the natural environment. However, it is difficult to imagine such a role for hope as much more than an ideal to which we must constantly aspire but never fully attain. I worry that hope is only a marker along the path to environmental action that orients us to a problem, and an illusion that covers up the inevitability of defeat. Yet, I myself hope we *know* this, but *act* as if the things for which we hope could in fact be effected in the world: in this way, we at least leave open the possibility that chance will prevail in sustaining the environments and communities on which we rely. I propose we follow advice similar to food journalist Michael Pollan's on what we should eat (Pollan 2007) and have hope, not too much, mostly for plants.

## NOTES

1. Part of this chapter was developed from a 2014 series of blog posts I authored as a postdoctoral scholar at Penn State University's Rock Ethics Institute.
2. This is substantial marker: Bill McKibben's efforts at 350.org exist to defend against surpassing what his organisation suggests as a 'safe' marker of 350 parts per million. (350.org).
3. For the distinction between *reflexivity* and *reflectivity*, see Beever and Brightman 2015.
4. For a thorough history of health-related conceptions of hope, see Jaklin Eliott's 2005 introduction to *Interdisciplinary Perspectives on Hope*.
5. See, for example, Mack et al. 2007, Miller 2012 and Patton et al. 2013.

*Part IV*

# Hope and Motivation

*Chapter Nine*

# To a Future Without Hope

Michael P. Nelson[1]

I don't hope for anything. I don't fear anything. I'm free.
—Nikos Kazantzakis[2]

We are bombarded with a contradiction. Many of our environmental leaders go to great lengths, masterfully and persuasively, to recount our many, looming, depressing and overwhelming environmental problems. We will suffer the various effects of massive global climate change, hundreds or thousands of species will be wiped from the face of this good green Earth and there will be tremendous and disproportionate human suffering along the path to this dimmer future. This is to say, our leaders convince us that the future is not only in jeopardy, but is essentially hopeless. These leaders then do something amazing, even audacious: they turn around and assert that there is hope, almost as a sort of unreflected-upon reflex, an utterance of seemingly obligatory expectation. I realize this may sound terrible, but this gesture toward hope has begun to make me angry. I am told a story convincing me there is no hope, and then I am told to have hope. You are not fooling anyone: to quote Leonard Cohen, 'Everybody knows that the ship is sinking, everybody knows that the captain lied'.

I worry that not only is hope a throwaway term—a vacuous sentiment we dole out at the end of a paragraph, a book, a film or an interview—but also that it is dangerous and counterproductive. To hope, quite literally, is 'to desire with expectation of obtainment' or 'to expect with confidence'. But the portrait our environmental leaders paint is one where there is no reasonable expectation of obtaining a desired end. Come on—it is either dishonest or lazy to tell me both that I cannot reasonably expect some future condition, and that I can confidently expect some future condition.

I am not saying that I dislike all uses or forms of hope. Sometimes I admit there seems to be so little at stake. Someone says or writes that we need to maintain hope in the face of great harms, great sorrows; the rest of us nod and resolve to do just that; and it is not clear that any of us—the speaker, the writer, the listener—really even knows what it is that we are hoping for, or what work hope is supposed to be doing. What really worries me—terrifies me, truth be told—is the use of hope as a motivator for healing our wounded and warped relationship with the natural world. I worry that hope will actually stifle, not aid, our resolve. I worry that hope can be, and often is, a distraction, an excuse for not getting on with the work at hand.

If you Google the words 'philosophy' and 'hope' together, one of the first hits is a cosmetics company, oddly called Philosophy, that peddles a skin moisturizer product even more oddly called 'Hope in a Jar'. Jarred hope is pricey—twenty-eight bucks for a single ounce. The company tagline reads, 'Where there is hope there can be faith. Where there is faith miracles can occur'. I have to fight my gag reflex here for the same reason that I fight it when I hear hope used as a motivator for action aimed at averting environmental harms.

There is another story of jarred hope: a story equally, though differently, sickening. In ancient Greek mythology, Pandora, sister-in-law of that fire thief Prometheus, was given a dowry and ordered by Zeus to keep it sealed. But curious Pandora (or her curious husband Epimetheus) could not resist. She opened the dowry and unleashed the scourges of humanity: greed, vanity, slander, envy, pining and other diseases. The last scourge, the last evil let loose, was hope ('hope' in Greek, Ελπις or *elpis*, is translated as 'anticipation of misfortune'). Even as early as 700 BC, Hesiod too expressed doubts about hope: 'Hope is a bad companion for the man in need who sits in an idle place, when he has no sufficient livelihood'.

I think I understand what is going on here. Our environmental leaders are looking to provide a motivation to act. And sometimes people act in desperation; sometimes they do amazing, creative and wondrous things. But sometimes, many times, they do not. I see it in my students nearly every day—the desire to do good, just and beautiful things in the world so easily quashed by the realization that what they decide to do won't make a difference, or that the world is beyond repair. Some other student (or professor!) flippantly points out that their individual decisions and actions do not make any real difference in the world, and the students are gutted, mouth agape, no response in sight. So they give up before they start. And I blame that on hope. As Francis Bacon once put it, 'Hope is a good breakfast, but a bad supper'. Hope is sugary cereal, quick yet vacuous energy for the masses—hope is not protein, meat that will suffice in a world gone awry.

How did we set this trap for ourselves? This, it seems to me, is the nub of the problem: hope results from a fixation on consequences—on judging right and wrong actions, picking professions and even justifying or evaluating the value of a lived life on what those actions, those professions and that life produce. We have built a society fixated on the future, perpetually risking all the attendant problems of justifying means by their ends and forever flirting with endorsing the hedonistic instincts of the masses. We have therefore built a society that can be readily disempowered.

It is time for a new form of motivation. Turning our backs on hope might be the best thing we can do at this moment in time. To be motivated by hope is to be stripped naked, to be vulnerable, to be disempowered. To be motivated by a sense of obligation, a commitment to virtue, is to put on a Kevlar bodysuit. 'Lack of power consists only in this', Baruch Spinoza points out, 'that a man allows himself to be guided by things outside him, and to be determined by them' (Spinoza 1992, 174 [Part IV, Prop. 37, Sch. 1]). Psychopathic serial killers tell us that the way they control their victims is to give them little tastes, little slivers of hope—but when their victims lose hope, they can no longer be controlled.

The writer Derrick Jensen nails it when he proclaims that he does not 'have much hope. But I think that's a good thing. Hope is what keeps us chained to the system, the conglomerate of people and ideas and ideals that is causing the destruction of the Earth. . . . A wonderful thing happens when you give up on hope, which is that you realize you never needed it in the first place' (Jensen 2006a). You are free to act rightly, because it is the right way to act and not because your action will move you or the world toward some future state.

So here's what I want, what I think we need so desperately. I want us to replace 'I hope' with 'I resolve to do the work' or 'I will be this kind of person, I will live this kind of life' or any sort of utterance that focuses on virtue rather than on consequence. Any sort of commitment that is not subject to the fickle and fragile focus on the results of our actions and commitments. This, I think, is the new ethic in the face of a future without hope. This is the only moral anchor imaginable in the sea change rolling our way.

I am calling for us to satisfy our obligation to the future by suggesting that we ignore, or at least greatly downplay, the consequences of our actions. I am suggesting instead that our obligation to the future is most properly satisfied when we act rightly and virtuously, and when our motivation stands stubbornly apart from, not held hostage to, the consequences of our actions.

If we are really going to resolve to act on behalf of the future, we need a motivation as steadfast and tough as the one Wendell Berry's character Dorie Catlett shows toward her forever-drunken uncle Peach. Referring to Dorie's moral resolve, Berry writes, 'she had long ago given up hope for uncle Peach. She cared for him without hope, because she had passed the place of turning back or looking back. Quietly, almost submissively, she propped herself against him, because in her fate and faith she was opposed to his ruin' (Berry 2004, 152).

What we need more than anything, need to save us from our own destructive fixation on the future, is no less than an individual and collective moral revolution: a revolution that includes abandoning hope, caring without hope and a commitment where we quietly, almost submissively, prop ourselves against those forces in the world that are working to bring ruin. We often hear that people only change their ideas, and therefore their behaviour, in the face of crisis. But we forget that a crisis can be a moral crisis as well, a sense of revulsion for a life that we are living, a commitment to live differently and to be a different kind of person. We need The Great 'Yuck!' Yuck, what we are doing is repulsive. Yuck, this is not the way a responsible person lives. The Great 'Yuck!' can be followed by The Great 'No!' No, I will not live this way. No, I will not be this kind of a person, this kind of an agent in the world. Finally, The Great 'No!' will give way to The Great 'Yes!' Yes, I will live a life of respect, of humility, empathy, care and attentiveness. Yes, I will choose to live with dignity and grace, no matter what. But none of this—the yucks or nos or yes'—is held hostage by the attainment of some future state. Each of us, right now, at this exact moment in time, has the power to choose to live the moral life, to live a life that is indeed worth living.

## NOTES

1. Reprinted courtesy of Trinity University Press. This appeared in the book *Moral Ground*, published by Trinity University Press. For more information, please visit www.tupress.org.
2. https://en.wikipedia.org/wiki/Nikos_Kazantzakis#/media/File:Kazantzakis_Grab.jpg.

## *Chapter Ten*

# Singing Hope's Praises

## *A Defense of the Virtue of Hope for Environmental Action*

Lisa Kretz

> Hope is the thing with feathers that perches in the soul—and sings the tunes without the words—and never stops at all.
> —Emily Dickinson [1]

I am interested in hope's role in action and specifically positive action regarding a moral harm I take to be unprecedented in human history, namely the ecological crisis. Only at this moment can we destroy with such force, efficiency and completeness. I take as given that the anthropocentric destruction of Mother Earth is morally wrong. My interest is in how to motivate and sustain behaviours that help preserve and advance the health of Earth's interrelated living members and systems. The focus in this chapter is on the potential that hope offers for motivating and sustaining human behaviours which positively impact the diverse entities and ecosystems that create the complexes of life as we know and value them on this planet.

I begin by defending the concept of hope that I adopt. Informed by the insights of psychology, hope has multiple manifestations, which include, but are not limited to, being a motivation for positive action and being a contagion. Moreover, hope can be taught, it enhances agency, hope is catchy, it has an abundance of beneficial outcomes, hope serves to empower and it is epistemically and socially grounded and responsive. I focus on the virtue of hope for positive ecological action as well as for personal well-being. Even with a multiplicity of beneficial outcomes to support the promulgation of hope, it is important to address potential counter-arguments. I explore worries about using hope as a tool for environmental action. These worries include the following: hope fails to adequately reflect the present, which can have negative results in terms of action; hope paints a picture so rosy that it may promote inaction; and a problematic ontology exists at the root of many accounts of hope where hope and despair are dichotomized and hope is taken to the be sole motivator. I also focus on the concern that hope is not strong enough to depend on for securing ecological action, and that its focus on consequences serves to undermine positive ecological action rather than ensure it. Once these potential problems have been addressed, I look at methodologies for teaching hope. These methods serve as an aid for teachers whose course content presents sufficiently daunting issues that the spectre of despair overshadows what could instead be an opportunity for a hopeful engagement with the world through empowerment and activism. Ultimately, I argue that hopefulness is an emotional and dispositional orientation that supports and enables positive ecological action.[2] Contrary to accounts that contend hope is neither a virtue nor does it play a beneficial role for those seeking ecological health, I argue that hope, when conceptualized through psychological findings, is clearly a virtue and that it plays a crucial role in action. If positive ecological action is the goal, we must nurture the virtue of hope, and I give concrete methods for doing so in the classroom setting.[3]

## CONCEPTUALIZING HOPE

On my understanding, an adequate conceptualization of hope will, of necessity, reflect the insights psychology offers. I am a practical moral philosopher in that I am not just interested in the 'why' but the 'how' of moral action. As such, looking at human psychology, including what it can reveal for helping people better reflect their considered moral beliefs, is essential to moral theory. I am interested in what emotions more generally have to offer the project of motivating moral action, but I concentrate on hope in what follows. Hope, on my account, is psychologically informed, motivates positive action, can be taught, enhances agency, is catchy, has a multitude of beneficial outcomes, empowers and is epistemically and socially responsive (Kretz 2013, 926, 929). I will defend this vision of hope in what follows. If I am correct, then there are a number of prima facie reasons for nurturing the virtue of hope.

I take hope, broadly construed, to pertain to a multidimensional and dynamic life force characterized by an orientation to a future good which remains uncertain; the hoped-for good is both realistically possible and personally significant and has implications for action and interpersonal relatedness (Dufault & Martocchio 1985, 350). The criterion that it must be realistically possible demands that hope must be justified; in the absence of justification, hope would be irrational given its groundlessness (Kretz 2013, 932). Indeed, a sufficiently correct appraisal of, and a sense of concern about, the current state of affairs are both necessary for generating hope with traction. It is because we are dissatisfied with the current state of affairs that we hope for change. Richard Lazarus, for example, suggests that a crucial 'condition of hope is that our current life circumstance is unsatisfactory—that is, it involves deprivation or is damaging or threatening' (1999, 654; Halpin 2001, 395).

To be justified, hope must be responsive to real-world constraints where we build on our knowledge and experience of past blocks to achieving our goals, as well as previous successes, and use this understanding to imagine ways forward to achieve further goals. Through setting smaller, reasonable, achievable, intermediate goals that contribute to larger goals, those larger goals become possible. There is a positive response when intermediate goals are reached which can help bolster energy for investment in additional goals (Macy & Johnstone 2012, 224). People who possess high levels of hope often prefer 'stretch goals' which are slightly more difficult than previously attained goals which facilitates likely success at intermediate goals and further growth in the direction of the long-term hoped-for goals (Marques et al. 2014, 36).

Probabilities matter when it comes to justified hope. The probability of goal attainment spans from a very high to very low probability, and justified hope requires reflection of these probabilities (Marques et al. 2014, 36). High hope thinking involves identifying specific goals as opposed to vague goals (Snyder 2002, 250). Stopping all ecological destruction is a lofty goal to be sure, but it is also vague. More concrete goals would involve identifying a specific form of ecological destruction and putting energy into working against the harm in ways that will be effective. Perhaps, for example, one seeks to minimize carbon dioxide emissions generated by transportation. To this end, one might choose to own a bike instead of a car and choose to purchase local and in-season produce. This immediately cuts down on your own carbon dioxide emissions, including carbon dioxide emissions associated with the transporting of nonlocal and out-of-season produce, thereby serving as a tangible and immediate move toward your overarching goal. For political influence, one might petition his or her local government for bike-friendly roadways, enhanced public transportation methods and increases in community-supported agriculture. To this end, one might also form a coalition for generating action around enhancing sustainable forms of transportation and food production and distribution. If a successful model is created over time for improved local commitments to reducing transportation-related carbon dioxide emissions, the model can be shared with other communities.

Hope, like many other emotions, is infectious; it can be transmitted like a contagion (Braithwaite 2004a, 11; Cunningham 2004, 9; Eliott 2005, 11). Studies show that hope can be taught; therefore, one method of transmitting hope is teaching. There is evidence of the successful teaching of hope, and resultant increases in hope, at the middle school, high school and university levels (Cheavens, Michael & Snyder 2005, 126). A nonexhaustive list of the virtues of increased hope includes positive influences on academic and athletic performance, student self-esteem, physical health, optimism, affectivity, expectancies of positive outcomes, perceived problem-solving capacities, perceptions of control in life and psychological adjustment (Cheavens, Michael & Snyder 2005, 119, 126; Snyder 1995, 357). It is also recognized as an essential motivator (Drahos 2004, 19). Those with higher hope have more mental energy for, and pathways to, their goals; when faced with blocks to their goals, they conceptualize alternative routes and possess the agentic thinking necessary for activating action (Snyder 1995, 357). High hope people have more goals and more difficult goals, which are perceived as challenges, and they have greater success at goal achievement, more happiness and less distress, superior coping skills, less self-reported work burnout and better recovery from physical injury (Snyder 1995, 357–58). Broadly, high hope is related to beneficial life outcomes, life satisfaction and well-being (Cheavens, Michael & Snyder 2005, 127; Marques et al. 2014, 38).

The agency associated with hope, in my view, is problematic if it is premised on an ontology of abstract individualism. Rather, the deeply social elements of self-construct and action need recognition for adequate approaches to achieving one's goals. Radical dependency on others for the majority of that which sustains our existence must be flagged, ranging from our ecological dependence on healthy water, air and food (Kretz 2009, 123–29) to social dependence on relationships with close and distant others (Young 2006). Communities of mutually responsive high hope persons generate a synergistic interanimation of others and one's own hopes, providing mutual supports that contribute to hope's energy (McGeer 2004, 109, 123). Networks of mutual support bring benefits such as higher levels of trust, enhanced health, lower suicide rates and less depression (Macy & Johnstone 2012, 123).

The version of hope I defend is a relational conceptualization of hope where hope is only possible in supportive relations. In the absence of such relations, hope is neither articulable (language, emotions and concepts are socially derived) nor justified (relations that support our projects are necessary for the achievement of goals). Darren Webb contends hope is a socially mediated human capability with varying behavioural, cognitive and affective dimensions (2013, 398). Achievement of individual hoped-for goals depend on wider circles of action by others (Drahos 2004, 20). Goal-directed and hopeful thinking is learned in the context of other people (Snyder 2002, 263). When individuals are moved by a shared vision, they become a community with a common purpose and what is achievable as a community surpasses that of individuals (Macy & Johnstone 2012, 163). Hope occurs in environments where a supportive atmosphere, meant to help meet individual and collective goals, grounds interactions; such environments enhance the perception that one has both the pathways and agency to succeed (Snyder 1995, 359). The relational dimensions of hope help to highlight the ways in which existing oppressive relations might undermine the development and sustenance of capacities for hope. It is essential to remedy blocks to the development of positive human capabilities. For example, neglect and physical abuse result in decreases in hope (Snyder 2002, 263). So too does witnessing family members or friends being subject to interpersonal violence (Marques et al. 2014, 40).

## ARGUMENTS IN FAVOUR OF NURTURING THE VIRTUE OF HOPE

### Action

There are a number of reasons, many which have already been articulated above, as to why one ought to nurture the virtue of hope. For those interested in positive ecological action, hope plays a necessary role in successful activism. What follows are pragmatic rationales for nurturing the virtue of hope if positive action is desired. Thus, one of hope's primary virtues is its motivating power for action.

First, at a basic level, hope is a prerequisite for action (Stotland 1969, 20–22). In the absolute absence of hope for successfully generating the result hoped for, there is no rational motivation for action. Examples of the importance of hope for action are gleaned by Ezra Stotland from schools, controlled laboratory experiments, concentration and prisoner-of-war camps, hospitals and peace-time disasters—all of which attest to 'the importance of hope for action, even action to prevent the greatest of all disasters, death' (Stotland 1969, 20–22). Having hope for the future is necessary for the continuation of life (Eliott 2005, 14; Farber in Snyder 1995, 358; Freire 2011, 2; Lazarus 1999, 659; McGeer 2004, 100; Stotland 1969, 8, 21–22). Although we may have more or less hope, healthy or unhealthy amounts, the requirement for hope in human beings is pervasive.

Hope is linked to action in that when you are hopeful you act in ways to bring about what is hoped for (Braithwaite 2004b; Cartwright 2004; Drahos 2004; Snyder 1995). My account of hope is strongly informed by that of Charles Snyder, who is a psychologist specializing in positive psychology. On Snyder's account, hope reflects 'individual's perceptions regarding their capacities to (1) clearly conceptualize goals, (2) develop the specific strategies to reach those goals (pathways thinking), and (3) initiate and sustain the motivation for using those strategies (agency thinking)' (Snyder et al. 2003, 122–23). Pathway and agency thinking are additive, iterative, reciprocal and positively related; both are necessary for sustaining successful goal pursuit and they are insufficient when they occur in isolation from each other (Snyder 2002, 252; Snyder et al. 2003, 123). The active element of hope is thereby highlighted. Pathway thinking increases the likelihood of goal achievement through conceptualizing multiple pathways to achieve goals when blocks are discovered, and agency thinking provides the sort of self- and community conceptualization that engenders success at achieving hoped-for goals. Colloquially put, agency thinking is 'the will' and the pathway thinking is 'the way' (Snyder 1995, 355). The power of one's agency is explored when planning produces action and the results of that action contribute to further planning (Drahos 2004, 22).

In Joanna Macy (a renowned deep ecologist) and Chris Johnstone's (a medical doctor specializing in the psychology of behaviour change) book *Active Hope: How to Face the Mess We're in Without Going Crazy*, they identify the formative role for hope in facing ecological crisis. When facing challenges, the kinds of responses we form, as well as the degree to which we believe they are relevant, are directly shaped by the way we think/feel about hope (Macy & Johnstone 2012, 2). Active hope, in their analysis, is a practice; it is something we do as opposed to something we just have (Macy & Johnstone 2012, 3). The narratives we tell ourselves about our past, current and future abilities shape the story to come. Macy & Johnstone encourage developing inner resources and community resources to enhance the strength of our capacity to face challenges and respond with resilience (2012, 6). Thus, the relevance of healthy community is highlighted. Ideal power is recognized as growing, of necessity, through and with healthy community. Ideal power structures are recognized as being collaborative; power is construed relationally as power-with rather than power-over (Macy & Johnstone 2012, 7, 109). One must first recognize the ecological harms taking place, for example, with regard to climate change, weather-related diseases, floods, droughts, and major hurricanes, to assess the state of affairs honestly and to correctly assess what changes need to be made (Macy & Johnstone 2012, 20). It is crucial to honour our pain for and with the world through a healthy response to the appropriate distress that is felt through exposure to large-scale and unnecessary destruction, as well as to express these feelings in ways that facilitate empowerment (Macy & Johnstone 2012, 66–67).

If positive ecological action is desired, then adopting a hopeful attitude which manifests as hopeful action is essential. Acting in hopeful ways serves to bring into being hoped-for goals that would be impossible in the absence of hopeful action. If you maintain the status quo, if you do not try to make things better, then the status quo will remain the case. If you try to improve the world, learn from your errors and build on your successes, then further related goals can be generated. If people truly believe an ecologically responsible society is not possible, this belief results in a sense of hopeless malaise where it is pointless to try, thereby making the belief that society cannot change a self-fulfilling prophecy (Milbrath 1995, 108, in Lueck 2007, 253). If, instead, people believe change is possible, they engage in the behaviours required for positive change, learn from failures and successes and build on that knowledge base for future action. Hope provides tools for plan creation and thereby aids in the planning, action, outcome and feedback loop that serves to generate or alter expectations and hopes (Lueck 2007, 252).

Hope is thus not only a precondition for action; it is a precondition for positive change. Future states of the world are dependent causally on the choice to hope or not to hope (Bovens 1999, 671). As I have argued elsewhere, there is a bootstrapping (hopeful action itself substantiates and generates further hopes) and snowballing (hope's catchiness generates increasing amounts of persons acting hopefully, which increases what groups can achieve and hope for) effect with hope (Kretz 2013, 940).

## Well-Being

Another reason for nurturing the virtue of hope is the impact it has on one's own experience of life. The absence of hope is despair, and despair carries with it the sense of being overcome by futility or defeat—it is to believe the possibility of getting the desired object or outcome does not exist (Govier 2011, 247). Paolo Freire takes despair and hopelessness to be 'the cause and consequence of inaction or immobilism' (2011, 3). Hopelessness involves disengagement; it is a release of agency where there is no belief that any action can make a difference, and as such no effort is made (Lueck 2007, 251). Educators and conservation psychologists worry that despair leads to terror management, and terror management is demonstrated through downplayed problems and hyper-materialism being used as ineffective coping mechanisms (Kelsey 2012). Lower hope scale scores for college students are related to suicidal ideation (Range and Penton 1994, in Snyder 2002, 261). High hope people encounter goal blocks with flexibility and find alternative goals, whereas low hope people ruminate about feeling stuck and engage in escape fantasies (Michael 2000, Snyder 1999, in Snyder 2002, 261). Such avoidance behaviour and disengaged coping have counterproductive consequences generally (Snyder and Pulvers 2001, Stanton and Snider 1993, in Snyder 2002, 261). Low hope persons continue in their passivity because they are preoccupied with avoidance thoughts and fail to learn from past experiences (Snyder 2002, 261). Low hope people also tend to be unforgiving of other people, fearful of interpersonal closeness and lonely (Thompson et al. 2002, in Snyder 2002, 262). Indeed, Snyder goes so far as to propose hope and meaning in life are companions; through self-reflection about personal goals and perceiving progress, life meaning is constructed (Snyder 2002, 262).

The personal narratives we generate have a significant impact on our sense of self-worth and what we are capable of achieving. High hope people tend to have an affirming and positive internal dialogue (Snyder et al. 1998, in Snyder 2002, 251). Agency thought, which is taken to be the motivational component in hope theory, refers to the perceived capacity to use pathways to reach desired goals (Snyder 2002, 251). High hope people embrace agency phrases such as 'I can do this', which play an essential role in navigating blocks to goal achievement (Snyder et al. 1998, in Snyder 2002, 251). Hope has been found to correlate positively with positive affect and negatively with negative affect (Snyder 2002, 261). Higher hope is associated with self-reports of fewer negative thoughts and increased positive thoughts (Snyder et al. 1996, in Snyder 2002, 261). Findings likewise confirm high hope generates feelings of confidence, inspiration, being energized, being challenged by life goals, elevated self-worth, satisfaction and low levels of depression (Chang 1998, Kwon 2000, Snyder et al. 1991, Snyder et al. 1996, Snyder et al. 1997, in Snyder 2002, 261). Hope is crucial for enhancing life quality as it generates thoughts and actions used to establish optimal functioning and satisfaction (Snyder 2002, 268). Nurturing the virtue of hope is thus a form of a self-kindness as well as kindness to others given the positive action hope produces. Serendipitously, in this case, enhanced personal well-being supports the probability of increased positive action, which also benefits the community.

## ARGUMENTS AGAINST HOPE

Although the advantages to nurturing hope are impressive, it is important to consider the potential downside to nurturing hope in the context of ecological crisis. In 'Abandon Hope', Michael Nelson and John Vucetich provide a series of key critiques which must be grappled with if hope theory is to play a central role in environmental theory and practice. They argue against construing hope in environmental ethics as a prime motivator of behaviours that generate critical mass and against conceptualizing hope as a fundamental virtue (Nelson and Vucetich 2009, 32–33). In particular, they are concerned about times in which hope for a sustainable future is taken to be the fundamental reason for changing one's relationship with nature (Nelson and Vucetich 2009, 33). Advocating an approach to remedying environmental harms primarily through appeal to a sole reason is destined to fail; no singular approach will, in isolation, meet the complex dimensions of anthropogenic environmental harms. Hoping, as an isolated approach, will be insufficient for ameliorating or removing ecological harms. Rather, multiple, creative, context-sensitive approaches will be required. But this is not to say that hope is, on principle, of concern. Nor is it to say that, on principle, hope cannot be a crucial ingredient for environmental action. Thus, I will focus on the particular construals of hope that concern Nelson and Vucetich and respond to each in turn.

## Delusional

To highlight their underlying worry about hope being perceived as a fundamental virtue, Nelson and Vucetich illustrate a circumstance where hope is deluded and has detrimental results. In the example given, a terminally ill person postpones mending broken relationships because she is hopeful that she will live; here hope is interpreted as believing in a certain outcome when instead the individual should have recognized this as an unfulfilled desire (Nelson and Vucetich 2009, 33). My conceptualization of hope does not fall subject to this criticism given the criterion of justifiability influencing the viability of hope. Justifiability is what, on my account, separates wishful thinking from hope. Contrary to the notion that hope encourages delusions about actual states of affairs, people who score high in the hope scale appear to calibrate goal expectations using relevant boundary conditions (Snyder 2002, 264). It is those with extremely low hope that evidence extreme delusions about reality, and these delusions interfere with attaining desired goals (Cramer and Dyrkacz 1998, Irving et al. 1990, in Snyder 2002, 265). Concern about the proposed virtue of hope will need to stem from a different source. The threat of inaction is just such a potential source.

## Promotes Inaction

Nelson and Vucetich recount the story of Pandora opening the jar that was to be her dowry, thereby releasing the scourges of humanity (greed, vanity, slander, envy, etc.); she shuts the jar, however, in time to keep hope contained therein (Nelson and Vucetich 2009, 32). It is postulated that hope might itself therefore be an evil (Nelson and Vucetich 2009, 32). Hope is often associated with naïve and superficial optimism as well as with women and children (Smith 2005, 47). In addition, it is often seen as quaint and religious (Johnson 2005, 48).[4] The association of hope with women, weakness and palliating properties finds its archetypical manifestation in the story of Pandora (Smith 2005, 59). Nelson worries that hope is a throwaway term, a vacuous sentiment, that is dangerous and counter-productive (Nelson 2011, 459). The hopefulness of women and children is often perceived as demonstrating a regressive disposition wherein fantasy predominates over reality (Smith 2005, 47). The unpalatable present is 'dealt with' through the avoidance hope provides via a projected, imagined future (Smith 2005, 47). Insofar as hope functions as a form of compensation, it 'lends itself to passivity and indifference towards instigating change. To the extent that the hoper relates to a given desirable outcome simply by waiting for it, hope becomes problematic not just from an ethical but also from a political point of view' (Smith 2005, 47). Nelson worries hope might stifle our resolve rather than aid it; he is concerned that it is often a distraction or an excuse for not doing the work that needs to be done (Nelson 2011, 459).

Again, such critique is not correctly launched at a conceptualization of hope that recognizes what psychology points to, namely the pathway and agency components of hope. Contrary to the concerns articulated above, rather than hope being associated with a naïve optimism, hope for a better state of affairs entails a critical response to the current state of affairs. Hoping to generate alternatives reflects discontent with present circumstances, and it implies the space and freedom for change (Halpin 2001, 395). It can be a useful tool for visualizing a state of affairs that doesn't yet exist, with anticipation helping to prepare the ground for what is yet to be (Halpin 2001, 395). Hope can serve to galvanize and redouble efforts when facing negative circumstances (Halpin 2001, 396). As opposed to hope being a threat to inaction, some postulate that despair frequently leads to a pessimism that contributes to a grudging acceptance of the status quo (Nesse 1999, in Halpin 2001, 396). Certainly hope as a psychological phenomenon instigates and sustains goal-oriented behaviour. Despair as a psychological phenomenon results in inaction. Sadly, when manifest as depression, it disproportionately results in permanent inaction through death at one's own hand.[5]

Moreover, the association of women with disempowering emotional proclivities merits its own analysis. However, given space constraints, I will only be able to gesture at this issue. There is a long history of feminist critique of the double bind many women and members of oppressed groups find themselves in when it comes to expressing emotion; it is simultaneously expected and used as a reason for ignoring the content of what one is attempting to express (Boler 1999; Campbell 1994, 1997; Frye 1983; Lloyd 1984; Spelman 1989).[6] The devaluation that emotions (including hope), femininity, women and members of oppressed groups face is not unrelated.

## Problematic Ontology

The ontology of hope to which I ascribe, where hope is precondition for positive action in the world, is potentially called into question by Vucetich and Nelson.[7] They recognize that, according to some theories, not being hopeful constitutes despair, and as such advocating action in the absence of hope might be unacceptable if not incomprehensible (Nelson and Vucetich 2009, 34). Contrary to this view, they question the hope/despair dichotomy being postulated as the sum of all ethical motivators (Nelson and Vucetich 2009, 35). I contend, however, acknowledging that hope motivates behaviour and despair does not, is not to say that these are the sum of all potential ethical motivators. There are a bevy of motivations for action, for example health, kindness, generosity and curiosity. However, if you had no hope that you could achieve health through your proposed actions, no hope that you could express kindness or generosity through your actions or no hope that your curiosity might be satisfied through exploration, then there would be lacking motivation to complete said action. Acting without hope for achieving your goal is irrational, and it certainly ought not to be promoted as a virtuous trait to nurture. This, however, worrisomely seems to be the direction that is gestured at by Nelson and Vucetich.

## Too Concerned with Consequences

Nelson and Vucetich worry about the hold that utilitarian philosophy continues to exert in the conceptualization of ethical problems and their solutions, and instead advocate a virtue approach. They worry that theorists of hope who focus on probabilities for particular outcomes of behaviours are overly committed to the consequentialism encapsulated in utilitarian approaches, and they worry about the focus on the future as opposed to the present that is implied by such approaches (Nelson and Vucetich 2009, 35). We often do not have control of the situations we are faced with, and we often cannot control outcomes; thus, we need a method for acting morally when consequences are beyond our control. When Al Gore's sister died of lung cancer, their father ceased farming tobacco, not because his actions would impact the future but because he saw it as the right, the virtuous, thing to do (Nelson and Vucetich 2009, 35). Extrapolating from this case, it is suggested that we should live in environmentally responsible and caring ways because such action is right; future effectiveness may be irrelevant or inappropriate as a consideration (Nelson and Vucetich 2009, 35). Sustainable living ought to be premised on manifesting basic virtues, recognized as good in and of themselves, as opposed to being premised on hope for a better future (Nelson and Vucetich 2009, 35). In such an analysis, whether critical mass for a better future is likely or possible is not the correct focus. Of course this need not be an either/or—either base your entire moral theory on attending to consequences or ignore them entirely[8]—however, the main point is that enacting environmentally virtuous behaviour regardless of outcomes potentially ensures moral behaviour in circumstances where uncontrollable consequences are on the horizon.

I question the proposed disconnect between virtuous action and consequences. Acting without taking into consideration the consequences would generate behaviour untethered to the world and what matters and why. Consequences are tied to contexts and contexts are tied to relationships. Aristotle, an esteemed virtue ethicist, notes that one must, for example, get angry at the right person, in the right amount, at the right time and for the right end

(Aristotle 1999, 29, 109a25). Context matters. The result of yelling at a very young child in anger is a distraught child. It results in the fear and upset of someone without the cognitive capacity to interpret why such negativity is being directed their way. In contrast, yelling 'Stop!' at someone hurting a child is appropriate in a way that yelling at a child is not. You need to be aware of the specific context, the nature of the relationships involved and the likely consequences of your action to decipher what constitutes virtuous action in relation with others (both human and nonhuman). What constitutes responsible and caring behaviour is discovered in reflexive relationship with environments and their members; as such, consequences directly inform what virtuous ethical behaviour entails. Although consequences are not the only things that need to be taken into account, they are certainly important things to take into account.

Nelson and Vucetich maintain that 'Living by such [basic] virtues is a fundamentally right way to live—even if nobody else does and even if it might not avert environmental disaster' (Nelson and Vucetich 2009, 35). I am fully in agreement that our inability to avert ecological disaster is not permission to treat others poorly, including environmental others.[9] But an inability to avert total eventual ecological destruction is not an inability to help the Earth members and systems that are present. When we are engaging in behaviours that will have no positive outcome for the environment, it may be a symbolically virtuous gesture, but insofar as it is completely futile I would hesitate to promote it as an unqualified virtue. Amassing futile actions would distract from actions that could actually help (Halpin 2001, 401). Part of the issue pertains to the example being used. Namely, the presumed impossibility of stopping the anthropocentric destruction of the planet as we know it given how far along in this destruction we are, and how massive the push is to continue in the vein is (e.g., rupturing the core of the Earth to extract a short-term fix to an energy addiction, otherwise known as fracking). As I have argued elsewhere, this is an error of scope (Kretz 2013, 936). To say we cannot end the demise of planet Earth as we know and love her is not to say that we cannot be as gentle with her, ourselves and others in the interim and that doing so is morally required. Although we all die, that does not mean our actions toward each other during the duration of our lives are rendered irrelevant because eventually they will not have the results intended. Insofar as we are tending to the present, there are concrete ways to help that will have concrete results. Furthermore, adequate attention to the present is precisely what is required for epistemically grounded hope.

Nelson worries that his students, who would otherwise do good deeds in the world, will fail to do so when they realize that what they do might not or will not make a long-term overarching difference—when they realize the world is beyond repair (2011, 460). This is a critical worry. Facilitating educational environments that foster empowerment and action is a necessary component for altered ecological behaviours. Nelson advocates adopting a stance wherein one acts in the absence of hope and instead derives their motivation through a sense of obligation and a commitment to virtue which is visualized as a Kevlar (DuPont, Wilmington, DE, USA) bodysuit compared to the naked, vulnerable, disempowering reliance on the motivation of hope (2011, 460).[10] He quotes approvingly of Baruch Spinoza claiming that lack of power consists solely in a man allowing 'himself to be guided by things outside him, and to be determined by them' (2011, 460). Derrick Jensen's stance in 'Beyond Hope' (2006a) is likewise embraced; giving up hope is a freeing occurrence wherein one recognizes what they ought to do (help the environment) regardless of what others do (harm the environment) (Nelson 2011, 461). Resolving to do the moral work necessary in the absence of hope is identified as a moral anchor in the sea change rolling our way (Nelson 2011, 461). Nelson suggests ignoring or at least downplaying the consequences of our actions; motivation is taken to stand stubbornly apart from the consequences of actions (2011, 461). My concerns about the separation between consequences and the moral rightness of an act were articulated above. I will now focus on the emotional disconnect being proposed.

Advocating action without hope and fortitude without the emotional content needed to keep it afloat fails to adequately reflect the role of emotion in motivation. Nelson recommends a moral revolution wherein one abandons hope and cares without hope (2011, 462). He advocates a robust recognition of how bad things actually are and then deciding to live with dignity and grace, with humility and empathy, and with care and attentiveness, regardless of what life presents us with (2011, 462). To illustrate the problem with this recommendation, let us consider an example with someone who fundamentally lacks hope. Telling someone who is clinically depressed, someone truly depleted of hope, to keep going through the motions indefinitely regardless of how he or she feels and with no legitimate hope that things will change for the better, at minimum fails to reflect the disempowering psychological results of despair and more strongly exhibits a lack of care. When one is that depleted of hope, what is needed is treatment. So too when activists are seeking to continue protesting ecological harm and feeling the pull of despair, they need therapy and connection to communities of hopeful people achieving change, not a demand for action without the required emotional, psychological and community support. When action without the requisite emotional support is advocated, there is a failure to recognize the necessary role particular emotions play in bringing to realization and sustaining action. Hope depletion does serious damage to the human psyche and results in decreased efficacy regarding action. A healthy psyche is needed for a realistic identification of the ecological challenges being faced and it is required for performing well in relationships. In other words, it is needed for performing virtuously.[11]

## TEACHING HOPE

Given the above defence of hope's virtues, I will now turn to methods for generating and supporting hope in classroom settings. An effective hope scale (for adults) and a children's hope scale (ages eight to sixteen) have been developed by Snyder and his colleagues to ascertain hope levels (Marques et al. 2014, 36–37; Snyder, Cheavens & Sympson 1997, 109). Such a tool is essential for measuring the efficacy of attempts to increase hope and generate the positive impacts of hope in individual and communal life. As was already noted, it has been confirmed that hope can be successfully taught at the middle school, high school and university level (Cheavens, Michael & Snyder 2005, 126). One such course can be found at the University of Wyoming, which has a college class aimed at teaching hopeful thinking, and it successfully raises student's levels of hope, as well as their self-esteem and academic performance (Curry et al. 1999, in Snyder 2002, 259). Hope interventions tend to benefit students with the lowest levels of hope the most; however, virtually all students experience an increase in hope levels through participating in school-administered hope programs (Marques et al. 2014, 41). Thus, there is a case to be made for making courses in hope available to all. There is also a case to be made for hope playing a role specifically in environmental classrooms where, given the scope of ecological harms, the threat of pessimism, despair and a general culture of doom and gloom loom large.

Cynicism and apathy are linked to civic disengagement, and pedagogical choices can either support or work against them (Johnson 2005, 44–46). If a class constantly focuses on problems, without adequate attention to solutions, then a culture of doom and gloom is created (Johnson 2005, 47). In addition, authoritarian teaching methods generate apathy through learned passivity, boredom and disengagement (Johnson 2005, 47). Connecting social problems to individual behaviours helps students to see that their behaviour does in fact have an impact, while the failure to do so obscures the impact individuals have on changing or merely replicating existing social and political structures (Johnson 2005, 49). To achieve the goal of empowering students in their abilities to contribute constructively to communities and larger society, Brett Johnson recommends that teachers nurture civic responsibility, increase perceived civic efficacy (which increases the likelihood of engagement in civic behaviour), develop civic skills and increase knowledge about society (which increases the likelihood of competence) (Johnson 2005, 49–50). A nonexhaustive list of potential engagement methods includes the following: doing readings in class that advocate social action, engaging in positive deviance, having activist guest speakers, directing student participation in civic groups, analysing the effectiveness of social movement organizations, lessening teacher cynicism, supporting student-centred active learning and student-led social action and engaging in service learning and community research (Johnson 2005, 50). Active learning strategies in particular might include role-playing activities, group presentations, internships, action research and advocacy projects (Hironimus-Wendt & Wallace 2009, 82).

A concrete example of an environmental course that included a unit on hope is found in Shih-Jang Hsu's study of an environmental education course focusing on investigation, evaluation, action training and fostering empowerment (Hsu 2004, 41). Students identify a local ecological issue and then generate, assess and execute their plan to contribute to remedying the issue. A unit on hope and empowerment, with stories of success and presentations by activists, is included to address the disempowering despair that might result from engagement with sizeable ecological harms (Hsu 2004, 41). Through pre-course and post-course testing, it was discovered that the course effectively generated a significant promotion in students' responsible environmental behaviour, environmental responsibility, intention to act, perception of knowledge of environmental issues, perception of knowledge of and skills for environmental action strategies and locus of control (Hsu 2004, 41).

There are a number of concrete steps that might be taken to generate hope in a classroom environment. Steps for enhancing hope in students involve the following (Marques et al. 2014, 42): (1) Administer the hope scale. (2) Educate students about hope theory and its relevance for positive outcomes. (3) Have students list important life components, determine priority and discuss satisfaction levels with each. This helps to give structure to hope for the student. (4) Work with students to create workable goals that are positive, specific, salient for the student, and attainable. Also help the student develop multiple pathways for goal achievement and flag agentic thought patterns for the goals set. (5) Have the student visualize and verbalize the steps needed to reach his or her goals. Identify, in collaboration with the student, the most effective pathways and methods for agency. (6) Students can report back on the process of goal attainment and, through collaboration, make adjustments and modifications to disparities in action or thinking that are potentially hindering the successful achievement of desired goals. (7) The above process is cyclical, requiring continuous assessment. Over time, and with increasing successes, the bulk of the responsibility for implementing hope theory in their unique life experiences can be assumed by the student (Marques et al. 2014, 42).[12] Hope is taught using such a methodology through practice and open dialogue about hope theory and its findings. Theory and practice therefore occur in tandem. I contend the behavioural 'doing' through engagement with the emotion of hope, as opposed to theoretical engagement alone, is essential to the success of teaching being hopeful (Kretz 2014).

Imparting hope to students rests on the foundation of helping them to set goals (Marques et al. 2014, 42). Breaking down goals into more manageable sub-goals is a useful strategy for engaging pathway thinking (Marques et al.

2014, 43). Through this practice, students can learn how to take a long-range goal and divide it up into manageable steps that lead to the end goal (Marques et al. 2014, 43). Helping students identify several routes to particular goals also is useful for helping students realize that potential blocks to success require revised plans as opposed to giving up outright (Marques et al. 2014, 43). It is also important to counter the negative, hypercritical, internal self-talk of low hope students through replacing perpetual self-criticism with realistic, positive and productive thinking (Marques et al. 2014, 44). Teachers can also engender hope in their students through developing and manifesting the characteristics of high hope individuals themselves (Marques et al. 2014, 44). The ripple effect of hope as a contagion in school settings can result in wider impacts (Marques et al. 2014, 46). The fact that teachers play a role in students' hope is evidenced by the fact that hope scale scores correlate significantly with a scale measuring teacher encouragement (Culver 1999 in Snyder 2002, 259). It is important, however, that the goals are the students' goals and not superimposed goals by their peers, parents or teachers (Marques et al. 2014, 43). When goals are imposed, they are not imbued with the same sense of motivation (Marques et al. 2014). It is crucial, then, when students are choosing action projects for achieving particular goals, that they be given the flexibility and support to choose actions and goals that are salient to them.

## CONCLUSION

I have argued in favour of a psychologically informed conceptualization of hope where hope garners justification through epistemic and social reflectivity and responsiveness. When the current state of affairs is unsatisfactory, we hope for change. Hope acts as a contagion and can be transmitted through teaching. It is both intrinsically and extrinsically desirable given the multitude of positive behaviours and dispositions it facilitates. Hope is, of necessity, relational and socially constituted.

I outline hope's fundamental role as a motivator for action. Hope theory identifies the capacities of high hoping persons which include clearly conceptualizing goals, developing pathways thinking (strategies) to reach those goals and agency thinking to initiate and sustain the motivation for utilizing those strategies (Snyder et al. 2003, 122–23). There is a snowball effect given hope's function as an emotive contagion and a bootstrapping that occurs when an initial orientation to hope makes it possible to achieve further goals related to what is hoped for. Additionally, a hopeful disposition supports general well-being.

A number of important worries were presented about hope's role in environmental action and were addressed in turn. I acquiesce that hope cannot be a stand-alone method for ameliorating ecological harm. No single approach will adequately address the complex dimensions of ecological harm that are currently occurring. Rather, multiple solutions engaging a variety of capacities will be necessary. To suggest, however, that hope is not a crucial ingredient for working against environmental harm fails to reflect the insights psychology offers. Contrary to the claim that hope encourages delusional thinking, statistically it is more likely low-hoping people will misread states of affairs than high-hoping people. For high-hoping people to generate appropriate goals and pathways, they must successfully read the actual state of the world. The feminization of hope and labelling it as weakness are likewise problematic for the sexism implied, for the devaluing of a vital emotion and for the failure to recognize hope's sturdy resilience. High hopers consistently generate new paths to their goals when faced with blocks to them, and they nurture the agency needed for fulfilling those goals. My conceptualization of hope is not subject to the criticism that it relies on an ontology where hope is the sole human motivator. I contend, rather, it is an element of motivated action but need not be the sole motivation. Last, I addressed the worries that hope generates too much concern with consequences and the claim that virtuous environmental action should happen regardless of the level of hope one has. First, I highlighted the necessary attention to consequences needed for correctly identifying virtuous behaviour. Second, I pointed to humans' psychological need for hope for action. Given the virtues of hope, coupled with its role in action, I outline a number of methods for cultivating hope in classroom settings.

I started this chapter with a quote by Emily Dickinson where hope is a bird singing a song that imbues life with music forever. It is a fitting metaphor given the role that birdsong, and its potential obliteration, played in helping the environmental movement come to fruition in the west with Rachel Carson's *Silent Spring* (1962). It is a comfort that I hear birdsong outside my window as I write this. A return to an Earth-responsive rhythm resonates with me as a method for bringing ecological hope and action into being. I imagine collective consciousness inspiring listening for the rhythm of the Earth beating heartfully. Such a vision requires intermediate and long-term goals happening across spans of individual and collective lives which will, given hope's power, generate positive gestalts not yet imagined. That is, if we nurture the virtue of hope.

## ACKNOWLEDGEMENTS

I wish to offer heartfelt thanks for the editorial recommendations of Joe Couture.

# NOTES

1. http://www.poetryfoundation.org/poem/171619.

2. My account includes a cognitive account of emotion though is not limited solely to a cognitive account.

3. I write from the perspective of an academic working in North America so my account of education is coloured and constrained by dominant North American educational models. Although I write from this perspective, and have been shaped by it, I do not contend these models are necessarily ideal.

4. In fact, Nicholas Smith hypothesizes that the most significant reason hope has generally been neglected as a topic of serious theoretical attention by philosophers and critical theorists is its association with religion. My account, in contrast, is a secular one.

5. Patients suffering from depression have a suicide rate at least several dozen times higher than that of the general population (Takahashi 2001, 360).

6. I say 'many' women to highlight the diversity of social and cultural expectations for differently situated women due to race, class, sexual orientation, gender manifestation, physical ability and so on. Intersectional forms of oppression importantly complicate any analysis if it is to be adequate.

7. Action here is to be understood as noncontingent and noncoerced.

8. This view, of course, runs contrary to the views of many deontologists and many consequentialists.

9. I have argued this elsewhere (Kretz 2013) in response to Andrew Fiala's (2010) worry in 'Nero's Fiddle: On Hope, Despair and the Ecological Crisis' that some may consider it rational to fiddle while Rome burns (a metaphor for the demise of life on our planet due to ecological harm).

10. Interestingly, the body suit in the analogy is used to protect the naked, vulnerable body underneath which taken to be hope. We protect what we value, thus hope's role in the analogy is evocative of its continuing value. My thanks to Michelle Willms for helpfully pointing this out to me in an earlier draft.

11. As Allen Thompson (2010) argues in 'Radical Hope for Living Well in a Warmer World', given the large-scale social, political and geographical shifts that will result from ecological harms that have already been set in motion, new virtues will come into existence. Developing a robust capacity for hope, which is both a virtue itself and an aid to the exercise of other virtues, will help ground and bring into being these new virtues.

12. The above recommendations were made with regard to students working with school psychologists. I leave it up to the reader to ascertain their transferability to the teacher-learner relationship between students and teachers.

# Bibliography

350.org. 'What We Do', 350.org. Accessed 10 May 2015. http://350.org/about/what-we-do/.

Abbasi, Daniel R., ed. 2005. *Americans and Climate Change: Closing the Gap Between Science and Action*. New Haven, CT: Yale School of Forestry and Environmental Studies.

Anderson, Kevin, and Alice Bows. 2008. 'Reframing the Climate Change Challenge in Light of Post-2000 Emission Trends', *Philosophical Transitions A* 366 (1882): 3863–3882.

Aristotle. 1999. *Nichomachean Ethics*, trans. Terence Irwin. Indianapolis: Hackett Publishing Company Inc.

Auden, W. H. 1976. *W. H. Auden: Collected Poems*, ed. Edward Mendelson. New York: Vintage International.

Barker, Dan. 2008. *Godless: How an Evangelical Preacher Became One of America's Leading Atheists*. Berkeley, CA: Ulysses Press.

Barnosky, Anthony, Nicholas Matzke, Susumu Tomiya, Guinevere O. U. Wogan, Brian Swartz, Tiago B. Quental, et al. 2011. 'Has the Earth's Sixth Mass Extinction Already Arrived?" *Nature* 471: 51–57.

Bebeau, Muriel, James Rest and C. M. Yamoor. 1985. 'Measuring Dental Students' Ethical Sensitivity'. *Journal of Dental Education* 49 (4): 225–35.

Beck, Ulrich. 2014. 'Positive Developments Can Result From Climate Change'. Accessed 1 July 2015. http://www.social-europe.eu/2014/03/climate-change/.

Beever, Jonathan. 2014. 'Hope and the Reset Button'. Accessed 1 June 2015. http://sites.psu.edu/rockblogs/2014/10/01/hoping-for-the-best-in-environmental-thought/.

Beever, Jonathan and Andrew Brightman. 2015. 'Reflexive Principlism as an Effective Approach for Developing Ethical Reasoning in Engineering'. *Science and Engineering Ethics* (epub ahead of print).

Berry, Wendell. 2004. *That Distant Land: The Collected Stories*. Berkeley, CA: Shoemaker & Hoard Press.

Black Elk and John G. Neihardt. 1979. *Black Elk Speaks: Being the Life Story of a Holy Man of the Oglala Sioux*, ed. John G. Neihardt. Lincoln: University of Nebraska Press.

Blohdorn, Ingolfur. 2002. 'Unsustainability as a Frame of Mind—and How We Disguise It: The Silent Counter-revolution and the Politics of Simulation'. *The Trumpeter* 18 (1): 1–11.

Boler, Megan. 1999. *Feeling Power: Emotions and Education*. New York: Routledge.

Bourget, David, and David Chalmers. 2014. 'What Do Philosophers Believe?' *Philosophical Studies* 170: 465–500.

Bovens, Luc. 1999. 'The Value of Hope'. *Philosophy and Phenomenological Research* 59 (3): 667–81.

Braithwaite, Valerie. 2004a. 'Collective Hope'. *Annals of the American Academy of Political and Social Science* 592: 6–15.

———. 2004b. 'The Hope Process and Social Inclusion'. *Annals of the American Academy of Political and Social Science* 592: 128–51.

Brand, Stewart. 2013. 'The Case for Reviving Extinct Species'. National Geographic News. Accessed 15 March 2015. http://longnow.org/revive/what-we-do/why-revive-extinct-species/.

Braungart, Michael, and William McDonough. 2002. *Cradle to Cradle*. New York: North Point Press.

Breyman, Steve. 1997. 'Social Studies of Science and Activism: STS as Campus Greening Movement'. *Philosophy and Social Action* 23 (1): 5.

Broome, John. 2004. *Weighing Lives*. Oxford: Oxford University Press.

Cahen, Harley. 1988. 'Against the Moral Considerability of Ecosystems'. *Environmental Ethics* 10 (3): 196–216.

Campbell, Sue. 1994. 'Being Dismissed: The Politics of Emotional Expression'. *Hypatia* 9: 46–65.

———. 1997. *Interpreting the Personal: Expression and the Formation of Feelings*. London: Cornell University Press.

Caputo, John D. 1997. *The Prayers and Tears of Jacques Derrida*. Bloomington, IN: Indiana University Press.

———. 2006. *The Weakness of God*. Bloomington, IN: Indiana University Press.

———. 2013. *The Insistence of God*. Bloomington, IN: Indiana University Press.

Carrington, Damian. 2014. 'Earth has Lost Half of its Wildlife in the Past 40 Years, says WWF'. *The Guardian*, 29 September. Accessed 30 April 2015. http://www.theguardian.com/environment/2014/sep/29/earth-lost-50-wildlife-in-40-years-wwf.

Carson, Rachel. 1962. *Silent Spring*. Boston: Houghton Mifflin.

Cartwright, John. 2004. 'From Aquinas to Zwelethemba: A Brief History of Hope'. *Annals of the American Academy of Political and Social Science* 592: 166–84.

Ceballos, Gerardo, Paul R. Erlich, Anthony D. Barnosky, Andrés García, Robert M. Pringle and Todd M. Palmer. 2015. 'Accelerated Modern Human–Induced Species Losses: Entering the Sixth Mass Extinction'. *Science Advances* 1 (5): e1400253.

Chawla, Louise, and Debra F. Cushing. 2009. 'Education for Strategic Environmental Behavior'. *Environmental Education Research* 13 (4): 437–52.

Cheavens, Jennifer, Scott Michael and Charles Snyder. 2005. 'The Correlates of Hope: Psychological and Physiological Benefits'. In *Interdisciplinary Perspectives on Hope*, ed. Jaklin Eliott. Hauppauge: Nova Science Publishers Inc., pp. 119–32.

Christensen, David. 2009. 'Disagreement as Evidence: The Epistemology of Controversy'. *Philosophy Compass* 4: 756–67.

Coeckelbergh, Mark, and Jessica Mesman. 2006. 'With Hope and Imagination: Imaginative Moral Decision-Making in Neonatal Intensive Care Units'. *Ethical Theory and Moral Practice* 10: 3–21.

Collingwood, Jane. 2015. 'The Relationship Between Mental and Physical Health'. *PsychCentral*. Accessed 20 April 2015. http://psychcentral.com/lib/the-relationship-between-mental-and-physical-health/0002949.

Crutzen, Paul J. 2002. 'Geology of Mankind'. *Nature* 415.

Cunningham, Kenneth. 2004. *An Autobiographical Approach to the Psychological Study of Hope*. Chicago: Institute of Clinical Social Work.

De Gere, Dawn. 2008. 'The Face of Masking: Examining Central Tendencies and Between-Person Variability in Display Management and Display Rules'. Dissertation, University of Washington.

Dewey, John. 1920. *Reconstruction in Philosophy*. New York: Henry Holt.

Dombrowski, Daniel. 1997. *Babies and Beasts: The Argument from Marginal Cases*. Champaign, IL: University of Illinois Press.

Drahos, Peter. 2004. 'Trading in Public Hope'. *Annals of the American Academy of Political and Social Science* 592: 18–38.

Dufault, Karen, and Benita Martocchio. 1985. 'Hope: Its Spheres and Dimensions'. *Nursing Clinics of North America* 20: 379–91.

Dweck, Carol. 2006. *Mindset: The New Psychology of Success*. New York: Random House.

Ehrenreich, Barbara. 2009. *Bright-Sided*. New York: Metropolitan Books.
Ehrlich, Paul, and Anne B. Ehrlich. 2013. 'Can a Collapse of Global Civilization Be Avoided?' *Proceedings of the Royal Society of Biological Sciences* 280: 2012–2845.
Eliott, Jacklin. 2005. 'What Have We Done with Hope? A Brief History'. In *Interdisciplinary Perspectives on Hope*, ed. Jaklin Eliott. Hauppauge: Nova Science Publishers Inc., pp. 3–45.
Ferguson, S. C., and E. A. Allen. 1880. *The Golden Gems of Life: Gathered Jewels for the Home Circle*. Cincinnati: Central Publishing House.
Fiala, Andrew. 2000. 'Toward an Ethics of Time: Eschatology and its Discontents'. *Philosophy in the Contemporary World* 7 (2–3): 33–41.
———. 2002. *The Philosopher's Voice*. Albany, NY: State University of New York Press.
———. 2009. 'How Would You Like to Be Him? The Golden Rule, Third Person Descriptions, and Virtue Ethics'. *The Pluralist* 4 (2): 24–37.
———. 2010. 'Nero's Fiddle: On Hope, Despair and the Ecological Crisis'. *Philosophy & the Environment* 15: 51–68.
Fisher, C. D., and N. M. Ashkanasy. 2000. 'The Emerging Role of Emotions in Work Life: An Introduction'. *Journal of Organizational Behavior* 21 (2): 123–29.
Fox, Warwick. 1995. *Toward a Transpersonal Ecology*. Albany: State University of New York Press.
Francis, Pope. 2015. 'On Care for Our Common Home'. Accessed 1 July 2015. http://w2.vatican.va/content/francesco/en/encyclicals/documents/papa-francesco_20150524_enciclica-laudato-si.html.
Frankl, Viktor. 2006. *Man's Search for Meaning*. Boston, MA: Beacon Press.
Freire, Paulo. 2011. *Pedagogy of Hope: Reliving Pedagogy of the Oppressed*. New York: Continuum Publishing Company.
Frye, Marilyn. 1983. 'A Note on Anger'. In *The Politics of Reality: Essays in Feminist Theory*. Trumansburg: The Crossing Press, pp. 84–94.
Gardiner, Stephen M. 2011. *A Perfect Moral Storm*. Oxford: Oxford University Press.
Garthoff, Jon. 2011. 'Meriting Concern and Meriting Respect'. *Journal of Ethics & Social Philosophy* 5: 1–28.
Ghosh, Pallab. 2015. 'Mammoth Genome Sequence Completed'. *BBC News*, 23 April. Accessed 15 May 2015. http://www.bbc.com/news/science-environment-32432693.
Gilbert, Jack. 2009. 'A Brief for the Defense'. In *Refusing Heaven*. New York: Knopf, p. 3.
Gosden, Emily. 2014. 'Look on Bright Side of Climate Change, Says IPCC Report Author'. Accessed 1 July 2015. http://www.telegraph.co.uk/earth/environment/climatechange/10733498/Look-on-bright-side-of-climate-change-says-IPCC-report-author.html.
Govier, Trudy. 2011. 'Hope and Its Opposites'. *Journal of Social Philosophy* 42: 239–53.
Gravlee, G. Scott. 2000. 'Aristotle on Hope'. *Journal of the History of Philosophy* 38 (4): 461–77.
Groopman, Jerome. 2005. *The Anatomy of Hope: How People Prevail in the Face of Illness*. New York: Random House.
Haack, Susan. 1997. 'The Ethics of Belief Reconsidered'. In *The Philosophy of Roderick M. Chisholm*, ed. Lewis Hahn. Peru, IL: Open Court, pp. 129–44.
Halpin, David. 2001. 'The Nature of Hope and Its Significance for Education'. *British Journal of Education Studies* 49: 392–410.
Hamilton, Clive. 2010. *Requiem for a Species*. New York: Routledge.
Hansen, James, Makiki Sato, Pushker Kharecha, David Beerling, Robert Berner and Valerie Masson-Delmotte. 2008. 'Target Atmospheric CO2: Where Should Humanity Aim?' *Open Atmospheric Sciences Journal* 2: 217–31.
Hardin, Garrett. 1968. 'The Tragedy of the Commons'. *Science* 162: 1243–48.
Held, Barbara S. 2002. 'The Tyranny of the Positive Attitude in America: Observation and Speculation'. *Journal of Clinical Psychology* 58 (9): 965–92.
Hironimus-Wendt, Robert, and Lora Ebert Wallace. 2009. 'The Sociological Imagination and Social Responsibility'. *Teaching Sociology* 37: 76–88.
Hoffman, Michael. 2015. 'In Violent Times, Young Japanese Just Shrug'. *Japan Times*, 7 February. Accessed 30 April 2015. http://www.japantimes.co.jp/news/2015/02/07/national/media-national/violent-times-young-japanese-just-shrug/.

Hsu, Shih-Jang. 2004. 'The Effects of an Environmental Education Program on Responsible Environmental Behavior and Associated Environmental Literacy Variables in Taiwanese College Students'. *The Journal of Environmental Education* 35: 37–48.

Hungerford, Harold R., and Trudy Volk. 1990. 'Changing Learner Behavior through Environmental Education'. *Journal of Environmental Education* 21 (3): 8–21.

James, William. 1890. *Principles of Psychology*. London: Macmillan.

———. 1956. *The Will to Believe*. New York: Dover Publications.

Jamieson, Dale. 2014. *Reason in a Dark Time: Why the Struggle Against Climate Change has Failed*. New York: Oxford University Press.

Jensen, Derrick. 2006a. 'Beyond Hope'. Accessed 9 September 2015. https://orionmagazine.org/article/beyond-hope/.

———. 2006b. *Endgame Volume 1: The Problem of Civilization*. New York: Seven Stories.

Johnson, Brett. 2005. 'Overcoming "Doom and Gloom": Empowering Students in Courses on Social Problems, Injustice, and Inequality'. *Teaching Sociology* 33: 44–58.

Kant, Immanuel. 1870. *Die Grundlegung zur Metaphysik der Sitten*. Berlin: L. Heiman.

———. 1959. *Foundations of the Metaphysics of Morals*, trans. L. W. Beck. Indianapolis: Bobbs-Merill.

Kaufman, Anthony. 2009. 'Climate Change in High Definition'. *Seed Magazine*, 22 April. Accessed 9 September 2015. http://seedmagazine.com/content/article/climate_change_in_high_definition/.

Keen, Sam. 1984. 'The Development of the Idea of Being'. In *The Philosophy of Gabriel Marcel*, ed. Paul Arthur Schilpp and Lewis Edwin Hahn. The Library of Living Philosophers, vol. 17. La Salle, IL: Open Court, pp. 99–120.

Kelsey, Elin. 2012. 'TEDxMonterey: Eco-buoyant!' Accessed 9 September 2015. http://www.youtube.com/watch?v=igVR2M6CMyM.

Kirchner, James W., and Anne Weil. 2000. 'Delayed Biological Recovery from Extinctions Throughout the Fossil Record'. *Nature* 404 (9): 177–80.

Kohák, Erazim. 1984. *The Embers and the Stars*. Chicago: University of Chicago Press.

Kretz, Lisa. 2009. 'Open Continuity'. *Ethics & the Environment* 14: 116–37.

———. 2013. 'Hope in Environmental Philosophy'. *Journal of Agricultural & Environmental Ethics* 26: 925–44.

———. 2014. 'Teaching Being Ethical'. *Teaching Ethics* 15: 151–72.

Kwon, Paul. 2002. 'Hope, Defense Mechanisms, and Adjustment: Implications for False Hope and Defensive Hopelessness'. *Journal of Personality* 70: 207–31.

Lazarus, Richard. 1999. 'Hope: An Emotion and a Vital Coping Resource Against Despair'. *Social Research* 66: 665–69.

Lear, Jonathan. 2006. *Radical Hope*. Cambridge, MA: Harvard University Press.

Leopold, Aldo. 1953. *Round River: From the Journals of Aldo Leopold*, ed. Luna B. Leopold. New York: Oxford University Press.

———. 1989. *A Sand County Almanac*. New York: Oxford University Press.

Levenbook, Barbara Baum. 1984. 'Harming Someone After His Death'. *Ethics* 94: 407–19.

Light, Andrew, and Holmes Rolston III. 2003. *Environmental Ethics: An Anthology*. Oxford: Blackwell.

Lloyd, Genevieve. 1984. *The Man of Reason: 'Male' and 'Female' in Western Philosophy*. Minneapolis: University of Minnesota Press.

Lobdell, William. 2009. *Losing My Religion: How I Lost My Faith Reporting on Religion in America*. New York: HarperCollins.

Lovelock, James. 2006. *The Revenge of Gaia.* New York: Basic Books.

———. 2008. 'Enjoy Life While You Can'." *Guardian*, 1 March 1st. Accessed 1 July 2015. http://www.guardian.co.uk/theguardian/2008/mar/01/scienceofclimatechange.climatechange.

Lueck, Michelle. 2007. 'Hope for a Cause as Cause for Hope: The Need for Hope in Environmental Sociology'. *The American Sociologist* 38: 250–61.

Mack, Jennifer, Joanne Wolfe, E. Francis Cook, Holcombe E. Grier, Paul D. Cleary and Jane C. Weeks. 2007. 'Hope and Prognostic Disclosure'. *Journal of Clinical Oncology* 25 (35): 5636–5642.

Macy, Joanna, and Chris Johnstone. 2012. *Active Hope: How to Face the Mess We're in Without Going Crazy*. Novato: New World Library.

Main, Douglas. 2013. 'Why Happy People Hide From Climate Change'. Accessed 1 July 2015. http://www.takepart.com/article/2013/05/22/happy-people-hide-from-climate-change.

Maniates, Michael, and John M. Mayer, eds. 2010. *The Environmental Politics of Sacrifice*. Cambridge, MA: MIT Press.

Marcel, Gabriel. 1949. *Position et approches concrètes du mystère ontologique*. Paris: Vrin.

———. 1950 and 1951. *The Mystery of Being*, vols. 1 and 2. London: The Harvill Press.

———. 1959. *Creative Fidelity*, trans. Robert Rosthal. New York: Farrar, Straus, and Co.

———. 1962. *Homo Viator*, trans. Emma Crawford. New York: Harper Torchbooks.

———. 1964. *Being and Having*. New York: Harper Torchbooks.

———. 1973. *Tragic Wisdom and Beyond*, trans. Stephen Jolin and Peter McCormick. Evanston, IL: Northwestern University Press.

———. 1995. 'On the Ontological Mystery'. In *The Philosophy of Existentialism*, trans. Manya Harari. New York: Citadel Press, pp. 9–46.

Marques, Susana, Shane Lopez, Sage Rose and Cecil Robinson. 2014. 'Measuring and Promoting Hope in Schoolchildren'. In *Handbook of Positive Psychology in Schools*, ed. Michael Furlong, Rich Gilman and Scott Huebner. New York: Routledge, pp. 35–50.

Martin, Adrienne. 2008. 'Hope and Exploitation'. *The Hastings Center Report* 38 (5): 49–55.

———. 2014. *How We Hope*. Princeton, NJ: Princeton University Press.

McGeer, Victoria. 2004. 'The Art of Good Hope'. *Annals of the American Academy of Political and Social Sciences* 592: 100–27.

McKibben, Bill. 1989. *The End of Nature*. New York: Random House.

———. 2006. *The End of Nature*. New York: Random House Trade.

———. 2007. *Deep Economy*. New York: Henry Holt.

———. 2010. *Eaarth: Making a Life on a Tough New Planet*. New York: Henry Holt.

Meeker, James. 1980. *The Comedy of Survival*. Los Angeles: Guild of Tutors.

Meirav, Ariel. 2009. 'The Nature of Hope'. *Ratio* 22 (2): 216–33.

Miller, Geoffrey. 2012. 'Hope is a Virtue'. *Journal of Child Neurology* 27 (12): 1616–17.

Montenegro, Maywa. 2009. 'Is There a Better Word for Doom?' *Seed Magazine* May 21.

Muir, John. 1979. *John of the Mountains: The Unpublished Journals of John Muir*, ed. Linnie Marsh Wolf. Madison: University of Wisconsin Press.

Naess, Arne. 1984. 'Identification as a Source of Deep Ecological Attitudes'. In *Deep Ecology*, ed. Michael Tobias. San Marcos, CA: Avant Books, pp. 256–270.

National Oceanic and Atmospheric Administration. 2015. 'Greenhouse Gas Benchmark Reached', 6 May. Accessed 15 May 2015. http://research.noaa.gov/News/NewsArchive/LatestNews/TabId/684/ArtMID/1768/ArticleID/11153/Greenhouse-gas-benchmark-reached-.aspx.

Nelson, Michael. 2011. 'To A Future Without Hope'. In *Moral Ground: Ethical Action for a Planet in Peril*. Trinity University Press: San Antonio, pp. 458–62.

Nelson, Michael, and John Vucetich. 2009. 'Abandon Hope'. *Ecologist* March: 32–35.

Nolt, John. 2006. 'The Move from *Good* to *Ought* in Environmental Ethics'. *Environmental Ethics* 28: 355–74.

———. 2009. 'The Move From *Is* to *Good* in Environmental Ethics'. *Environmental Ethics* 31 (2): 135–54.

———. 2010. 'Hope, Self-Transcendence and Environmental Ethics'. *Inquiry* 53: 162–82.

———. 2013. 'Anthropocentrism and Egoism'. *Environmental Values* 22 (4): 441–59.

———. 2015. *Environmental Ethics for the Long Term: An Introduction*. New York: Routledge.

Norcross, Alistair. 2004. 'Puppies, Pigs, and People: Eating Meat and Marginal Cases'. *Philosophical Perspectives* 18: 229–45.

Norgaard, Karie Marie. 2006. '"People Want to Protect Themselves a Little Bit": Emotions, Denial, and Social Movement Nonparticipation'. *Sociological Inquiry* 76 (3): 372–96.

Norem, Julie K., and Edward C. Chang. 2002. 'The Positive Psychology of Negative Thinking'. *Journal of Clinical Psychology* 58 (9): 993–1001.

Noss, Reed F., Allen Y. Cooperrider, and Roger Schlickeisen. 1994. *Saving Nature's Legacy: Protecting and Restoring Biodiversity*. Washington, DC: Island Press.

Oettingen, Gabriele. 2014. *Rethinking Positive Thinking*. New York: Current.

Ok, Uzeyir. 2004. 'Handling Doubt in Teaching Religion: A Turkish Case Study'. *Teaching Theology and Religion* 7 (4): 201–12.

O'Neill, John. 1992. 'The Varieties of Intrinsic Value'. *Monist* 75 (2): 119–37.

———. 1993. 'Future Generations; Present Harms'. *Philosophy* 68 (263): 35–51.

*The Onion*. 2014. '"It's Not too Late to Reverse the Alarming Trend of Climate Change": Scientists Who Know It's Too Late Announce'. Accessed 9 September 2015. http://www.theonion.com/articles/its-not-too-late-to-reverse-the-alarming-trend-of,34896/.

Orr, David. 2004. 'Hope in Hard Times'. *Conservation Biology* 18 (2): 295–98.

Palmer, C. Eddie. 1983. 'A Note About Paramedics' Strategies for Dealing with Death and Dying'. *Journal of Occupational Psychology* 12: 83–86.

Partridge, Ernest. 1981. *Responsibilities to Future Generations*. Buffalo: Prometheus Books.

Patton, D. J., R. M. Busch, K. M. Yee, C. S. Kubu, J. Gonzalez-Martinez and P. J. Ford. 2013. 'Hope Language in Patients Undergoing Epilepsy Surgery'. *Epilepsy and Behavior* 29: 90–95.

Pepper, David. 2005. 'Utopianism and Environmentalism'. *Environmental Politics* 14 (1): 3–22.

Pettit, Phillip. 2004. 'Hope and Its Place in Mind'. *Annals of the American Academy of Political and Social Science* 592.

Pinker, Steven. 2011. *The Better Angels of Our Nature: Why Violence Has Declined*. New York: Penguin Books.

Pojman, Louis. 1997. *Environmental Ethics: Readings and Theory and Application*. Belmont, CA: Wadsworth.

Pollan, Michael. 2007. 'Unhappy Meals'. *The New York Times Magazine*, 28 January. Accessed 15 June 2015. http://michaelpollan.com/articles-archive/unhappy-meals/.

Public Religion Research Institute and American Academy of Religion. 2014. *Believers, Sympathizers, & Skeptics: Why Americans are Conflicted about Climate Change, Environmental Policy, and Science*. Accessed 9 September 2015.http://publicreligion.org/site/wp-content/uploads/2014/11/2014-Climate-Change-FINAL1.pdf.

Regan, Tom. 1989. 'The Case for Animal Rights'. In *Animal Rights and Human Obligations* (2nd ed.), ed. Tom Regan and Peter Singer. Englewood Cliffs, NJ: Prentice Hall, pp. 13–26.

———. 2004. *The Case for Animal Rights*, updated. Berkeley: University of California Press.

Remen, Rachel Naomi. 1997. *Kitchen Table Stories*. New York: Riverhead.

Riffkin, Rebecca. 2014. 'Climate Change Not a Top Worry in the U.S.'. *Gallup*. Accessed 9 September 2015.http://www.gallup.com/poll/167843/climate-change-not-top-worry.aspx.

Roberts, Richard H. 1987. 'Review Article: An Introductory Reading of Ernst Bloch's *The Principle of Hope*'. *Journal of Literature and Theology* 1 (1): 89–112.

Rolston, Holmes III. 1988. *Environmental Ethics: Duties to and Values in the Natural World*. Philadelphia: Temple University Press.

Romm, Joseph. 2007. 'Debunking Shellenberger & Nordhaus Part II'. *Climate Progress*. Accessed 9 September 2015. http://grist.org/article/debunking-shellenberger-nordhaus-part-ii/.

St. Exupéry, Antoine de. 1967. *Wind, Sand, and Stars*, trans. Lewis Galantiére. New York: Harvest Books.

Scruton, Roger. 2013. *The Uses of Pessimism*. Oxford: Oxford University Press.

Seligman, Martin. 2006. *Learned Optimism*. New York: Vintage Books.

Shade, Pat. 2001. *Habits of Hope*. Nashville: Vanderbilt University Press.

Şhafak, Elif. 2010. *The Forty Rules of Love: A Novel of Rumi*. New York: Viking Penguin.

Shaw, Charles. 2006. 'The Priest and the Prophet: Can Industrial Civilization be Sustainable? Should It?' *Grist Magazine*, 15 August. Accessed 9 September 2015. http://grist.org/article/shaw1/.

Sheppard, James W. 2004. 'Reducing Pessimism's Sway in the Environmental Ethics Classroom'. *Worldviews* 8 (2–3): 213–26.

Shiu, Eric. 2014. *Creativity Research: An Interdisciplinary and Multidisciplinary Research Handbook*. New York: Routledge.

Simpson, Christy. 2004. 'When Hope Makes Us Vulnerable: A Discussion of Patient-Healthcare Provider Interactions in the Context of Hope'. *Bioethics* 18 (5): 428–47.

Singer, Peter. 1990. *Animal Liberation*, revised edition. New York: Avon Books.

———. 2002. *Animal Liberation*. New York: HarperCollins.

———. 2009. 'Speciesism and Moral Status'. *Metaphilosophy* 40: 567–81.

Smith, Nicholas. 2005. 'Hope and Critical Theory'. *Critical Horizons* 6: 45–61.

Snow, Nancy E. 2013. 'Hope as an Intellectual Virtue'. In *Virtues in Action: New Essays in Applied Virtue Ethics*, ed. Michael W. Austin. New York: Palgrave MacMillan, pp. 152–170.

Snyder, Charles. 1995. 'Conceptualizing, Measuring, and Nurturing Hope'. *Journal of Counseling & Development* 73: 355–60.

———. 2002. 'Hope Theory: Rainbows in the Mind'. *Psychological Inquiry* 13 (4): 249–75.

Snyder, Charles, Jennifer Cheavens and Susie Sympson. 1997. 'Hope: An Individual Motive for Social Commerce'. *Group Dynamics: Theory, Research, and Practice* 1: 107–18.

———. 2002. 'Hope Theory: Rainbows in the Mind'. *Psychological Inquiry* 13: 249–75.

Snyder, Charles, Shane Lopez, Hal Shorey, Kevin Rand and David Feldman. 2003. 'Hope Theory, Measurements, and Applications to School Psychology'. *School Psychology Quarterly* 18 (2): 122–39.

Snyder, Gary. 1990. *The Practice of the Wild*. Berkeley, CA: Counterpoint.

Spelman, Elizabeth. 1989. 'Anger and Insubordination'. In *Women, Knowledge, and Reality*, ed. Ann Garry and Marilyn Pearsall. Winchester: Unwin Hyman, pp. 263–73.

Spinoza, Baruch. 1992. *Ethics: Treatise on the Emendation of the Intellect and Selected Letters*. Indianapolis, IN: Hackett.

Stoknes, Per Espen. 2015a. 'The Coming Climate Disruptions: Are You Hopeful?' *Psychology Today*, 6 April. Accessed 7 April 2015. https://www.psychologytoday.com/blog/what-we-think-about-global-warming/201504/the-coming-climate-disruptions-are-you-hopeful.

———. 2015b. *What We Think About When We Try Not To Think About Global Warming*. White River Junction, VT: Chelsea Green Publishing.

Stotland, Ezra. 1969. *The Psychology of Hope: An Integration of Experimental, Clinical and Social Approaches*. San Francisco: Jossey Bass Inc.

Takahashi, Yoshitomo. 2001. 'Depression and Suicide'. *Japan Medical Association Journal* 44: 359–63.

Taylor, Paul. 1986. *Respect for Nature: A Theory of Environmental Ethics*. Princeton, NJ: Princeton University Press.

Texas A&M University. 2005. 'Humor Can Increase Hope, Research Shows'. *Science Daily*. Accessed 9 September 2015. http://www.sciencedaily.com/releases/2005/02/050211095658.htm.

Thompson, Allen. 2010. 'Radical Hope for Living Well in a Warmer World'. *Journal of Agriculture and Environmental Philosophy* 23: 43–59.

Thoreau, H. D. 1981. 'Walden'. In *Walden and Other Writings*, ed. Joseph Wood Krutch. New York: Bantam, p. 111.

Tipler, Frank. 1994. *The Physics of Immortality*. New York: Doubleday.

Tobias, M., ed. 1984. *Deep Ecology*. San Marcos, CA: Avant Books.

Tolstoy, L. 1978. 'A Confession', Section IV. In *The Portable Tolstoy*, ed. J. Bayley, trans. A. Maude. New York: Viking Penguin, pp. 666–731.

Tong, Rosemarie. 1993. *Feminine and Feminist Ethics*. Belmont, CA: Wadsworth.

Treanor, Brian. 2010. 'Blessed are Those Who Have Not Seen and Yet Believe'. *Analecta Hermeneutica* 2.

———. 2014. *Emplotting Virtue*. New York: Fordham University Press.

Treanor, Brian, Bruce Ellis Benson and Norman Wirzba, eds. 2015. *Being-in-Creation: The Human Place in the Natural World*. New York: Fordham University Press.

Tuana, Nancy. 2007. 'Conceptualizing Moral Literacy'. *Journal of Educational Administration* 45 (4): 364–78.

———. 2014. 'An Ethical Leadership Developmental Framework'. In *The Handbook of Ethical Educational Leadership*, eds. C. Branson and E. Gross. New York: Routledge, pp. 153–75.

Van Vechten, Diana. 2013. 'Impact of Home Literacy Environments on Students from Low Socioeconomic Status Backgrounds'. *Education Masters*. Thesis 248.

Verdenius, Willem Jacob. 1985. *A Commentary on Hesiod: Works and Days, Vv. 1-382*. The Netherlands: Brill.

Vetter, Craig. 1997. 'He's Not Worthy: A Portrait of a Millionaire at a Crux'. *Outside Magazine*, January.

Verghese, Abraham. 2004. 'Hope and Clarity: Is Optimism a Cure?' *New York Times Magazine*, February 22.

Von Scheve, Christian. 2012. 'Emotion Regulation and Emotion Work: Two Sides of the Same Coin?'. *Frontiers in Psychology* 3: 496.

Warner, Douglass, and David DeCosse. 2009. 'Environmental Virtue Ethics'. Accessed 1 July 2015. http://www.scu.edu/ethics/practicing/focusareas/environmental_ethics/lesson6.html.

Warraq, Ibn. 1995. *Why I Am Not a Muslim*. New York: Prometheus Books.

Webb, Darren. 2013. 'Pedagogies of Hope'. *Studies in Philosophy and Education* 32: 397–414.

Weber, Max. 1993. *The Sociology of Religion*. Boston, MA: Beacon Press.

Williston, Byron. 2012. 'Climate Change and Radical Hope'. *Ethics and the Environment* 17 (2): 165–86.

Wilson, E. O. 1986. *Biophilia*. Cambridge, MA: Harvard University Press.

———. 1994. *Naturalist*. Washington, DC: Warner Books.

———. 2010. *Anthill*. New York: Norton.

Ying, Y. W. 2007. 'The Buffering Effect of Self-detachment against Emotional Exhaustion among Social Work Students'. *Journal of Religion & Spirituality in Social Work* 27 (1/2): 127–46.

Young, Iris Marion. 2006. 'Responsibility and Global Justice: A Social Connection Model'. *Social Philosophy and Policy* 23: 102–30.

# Index

# About the Contributors

**Elizabeth Andre**, Ph.D., is an Associate Professor of Outdoor and Environmental Education at Northland College, an environmental liberal arts school in Ashland, Wisconsin, where she teaches courses on the societal response to climate change. Previously, she was the education coordinator and curriculum developer for the Will Steger Foundation's *Global Warming 101* initiative.

**Jonathan Beever**, Ph.D., is an Assistant Professor of Philosophy and faculty with the Texts & Technology Program at the University of Central Florida. He has held postdoctoral positions with Penn State's Rock Ethics Institute and with Purdue University's Weldon School of Biomedical Engineering. He has held fellowships with the Kaufmann Foundation, the Aldo Leopold Foundation and the Global Sustainable Soundscape Network. Beever works and publishes at the intersection of environmental ethics and bioethics, focusing on questions of ethics and representation. He teaches a wide variety of undergraduate and graduate courses on related topics.

**Andrew Brei**, Ph.D., is a Visiting Assistant Professor of Philosophy at St. Mary's University in San Antonio, Texas. His research has mostly to do with the ways in which human rights, health and psychology relate to the natural environment. He is regularly invited to speak to environmental science classes about the moral components of sustainability and about Aldo Leopold's Land Ethic. Brei's work has been published in the *Journal of Agricultural and Environmental Ethics* and the *Southwest Philosophical Review*. He teaches courses on ethics, the history of Western philosophy and philosophical anthropology.

**Andrew Fiala**, Ph.D., is Professor of Philosophy, Chair of the Philosophy Department and Director of the Ethics Center at California State University, Fresno. He is the author of a number of books and articles on ethics, war and peace, religion and political philosophy. Recent books include *The Peace of Nature and the Nature of Peace* (2015), *The Bloomsbury Companion to Political Philosophy* (2015), *Civility and Education in a World of Religious Pluralism*, co-edited with Vincent Biondo (2013) and *Ethics: Theory and Contemporary Issues*, eighth edition, with co-author Barbara MacKinnon (2014). Fiala also writes a regular column on religion and ethics for the *Fresno Bee*.

**Trevor Hedberg** is a Ph.D. Candidate and Graduate Teaching Associate in the Philosophy Department at the University of Tennessee. He is currently writing a dissertation on population ethics, and his general research interests include ethics, environmental ethics and epistemology. He has published articles in *Environmental Values*, *Synthese* and the *Journal of Business Ethics*. The Association for Practical and Professional Ethics selected him as the recipient of the Karen M. T. Muskovitch Award for Graduate Work in Professional Ethics in 2012.

**Lisa Kretz**, Ph.D., is an Assistant Professor of Philosophy at the University of Evansville. Her main areas of research are ethics (with a special emphasis on environmental and nonhuman animal ethics), critical pedagogy, student empowerment, the theory-action gap, moral psychology and anti-oppression activism. Her articles have appeared in the *Journal of Agricultural & Environmental Ethics*, the *Journal for Critical Animal Studies*, *Ethics & the Environment* and *Ethics and Education*.

**Michael P. Nelson**, Ph.D., is an environmental scholar, writer, teacher, speaker, consultant and professor of environmental ethics and philosophy. He holds the Ruth H. Spaniol Chair of Renewable Resources and serves as the Lead Principal Investigator for the HJ Andrews Experimental Forest Long-Term Ecological Research program at Oregon State University. He is the Philosopher in Residence of the Isle Royale Wolf-Moose Project, the longest continuous study of a predator-prey system in the world. Michael is the co-founder and co-director of the Conservation Ethics Group, an award-winning environmental ethics consultancy group fusing ethics with social and ecological science, and serves as a senior fellow for the Spring Creek Project for Ideas, Nature and the Written Word.

**John Nolt**, Ph.D., is Professor of Philosophy at the University of Tennessee. His research is centred on philosophical logic and environmental and intergenerational ethics, areas in which he has published seven books—most recently *Environmental Ethics for the Long Term: An Introduction* (2015). A long-time environmental activist, he has for the past decade chaired the University of Tennessee's Committee on the Campus Environment, which advises the administration on environmental matters. He was President of the University of Tennessee Faculty Senate in 2008–2009, and President of the state-wide Tennessee University Faculty Senates in 2009–2010. Much of his recent work is on climate and energy ethics.

**Brian Treanor**, Ph.D., is Charles S. Casassa Chair and Professor of Philosophy at Loyola Marymount University. He is the author of *Aspects of Alterity* (2006) and *Emplotting Virtue* (2014), and the co-editor of *A Passion for the Possible* (2010), *Interpreting Nature* (2013), *Being in Creation* (2015) and *Carnal Hermeneutics* (2015). His current projects include a monograph developing an 'earthy' hermeneutics engaged with the carnal body and the material world, and another on the experience of joy in the midst of finitude.